Science Pitch

Springer Nature More Media App

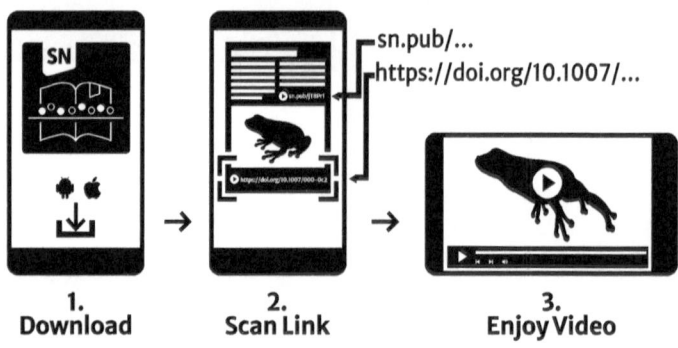

sn.pub/...
https://doi.org/10.1007/...

1.
Download

2.
Scan Link

3.
Enjoy Video

Support: customerservice@springernature.com

Stephen Wagner

Science Pitch

Present your Research.
Get to the Point

 Springer VS

Stephen Wagner
Inhaber
Redelandschaften.de
Bonn, Nordrhein-Westfalen, Germany

ISBN 978-3-658-44843-1   ISBN 978-3-658-44844-8   (eBook)
https://doi.org/10.1007/978-3-658-44844-8

Editorial Contact: Barbara Emig-Roller

This Springer VS imprint is published by the registered company Springer
Fachmedien Wiesbaden GmbH, part of Springer Nature.
The registered company address is: Abraham-Lincoln-Str. 46, 65189 Wiesbaden,
Germany

If disposing of this product, please recycle the paper.

Contents

Welcome: What You Can Take Away from This Book

1

Abstract

In this book, you will learn how to present your project precisely and convince your audience so that your research receives funding. Optimal preparation helps you avoid typical pitfalls and gives your target audience the most significant possible added value. You will learn how to strategically use artificial intelligence and familiarize yourself with the optimal setting for your presentation.

With the Science Pitch Canvas, you gather and prioritize essential aspects for your presentation. You integrate professional expertise and personality into a profound, brief presentation. This book helps you develop practice-oriented Science Pitches that you can flexibly adapt to your individual goals and for a specific audience. This way, you present your project to the point, arouse the interest of your audience, and expand your network. You will also be able to impress in interviews and secure funding for your research project.

An appealing presentation of your research that suits the audience is part of the standard professional qualification. But how can you present your project with power to the point? How do you convince and inspire your audience? And how do you elaborate the essentials of your project so precisely that your audience is convinced it is worthy of funding?

© The Author(s), under exclusive license to Springer Fachmedien Wiesbaden GmbH, part of Springer Nature 2024
S. Wagner, *Science Pitch*,
https://doi.org/10.1007/978-3-658-44844-8_1

This book provides all the essential information for delivering a first-class Science Pitch. It is divided into three parts.

For optimum preparation, we dive into the definition and meaning of Science Pitches. You will become familiar with your audience long before your presentation. You will learn how to use artificial intelligence (AI) as a sparring partner before we talk about common pitfalls for presentations: how much personality is appropriate? What content do you prioritize? How will your audience understand you? What is left out unnecessarily? Using specific examples, you will learn how to use power to get to the point when it comes to visual communication of essentials. Finally, we will prepare the ideal setting for your presentation.

In the second section, you create your Science Pitch Canvas. This will provide a quick overview of the essential aspects of your short presentation. It is about your technical expertise and how you can embed scientifically sound content of your project in professional storytelling. We illustrate why an authentic appearance and personal enthusiasm are essential for a convincing performance. You will build a bridge to your audience by emphasizing your project's relevance and innovative character. Not least, it is all about the essence of your project and a meaningful, memorable take home message.

The third part is dedicated to the practice: you will use specific examples to create and deliver your Science Pitch. For better orientation, you will find a practical example of a Science Pitch by a PhD student. Depending on the specific situation, you can adapt your Science Pitch to successfully present your project in 3 min, in less than a minute, or even in longer presentations. This will spark the interest of your audience for further information. You will enlarge your network and keep in touch with the people of your target group.

Doing so will align your personal goals with your audience's interests. You will be able to present a convincing Science Pitch, expand your network in personal discussions, and disclose your personality and professional expertise in interviews straight to the point. This way, you will fulfill an important prerequisite for successfully approving research funding.

I wish you lots of success in creating and presenting your Science Pitch!

Part I

Prepare Optimally for Your Science Pitch

Before you develop and articulate the content of your short presentation, we take a side trip into your setting: we clarify what the Science Pitch stands for, what your audience expects, and what prior knowledge they already have. We look at typical pitfalls that make well-intentioned presentations look bad. This section will prepare you to deliver an outstanding performance compared to many average speakers.

The Crucial Difference Between Publication and Presentation

Abstract

The standard format IMRAD (**I**ntroduction—**M**ethods—**R**esults **a**nd **D**iscussion), known for scientific publications, limits speech creativity and flexibility. Instead, presentations need personality and character to be memorable. A convincing Science Pitch consists of surprising twists of professionally substantiated content that matches the speaker's personality and passion. Link the content of your project into a compelling narrative or storyline. This way, you inspire and excite your audience and advance your career successfully. A prime example of this is the Science Pitch by Dr. Mai Thi Nguyen-Kim.

Do you know IMRAD? This standardized structure for publishing scientific papers represents the sequence **I**ntroduction—**M**ethods—**R**esults **a**nd **D**iscussion. This may be appropriate in the context of publications but does not fit an energetic presentation. The predictable IMRAD structure makes the presentation seem dull. In addition, this corset leaves little room for a creative and original presentation with surprising twists and turns. A personal, inspiring presentation is more impactful and memorable for your audience.

© The Author(s), under exclusive license to Springer Fachmedien Wiesbaden GmbH, part of Springer Nature 2024
S. Wagner, *Science Pitch*,
https://doi.org/10.1007/978-3-658-44844-8_2

So, what makes a Science Pitch? It needs an introduction that immediately engages your audience; then, there is a concrete solution or new insight. You may incorporate parts of IMRAD into your Science Pitch. But avoid its strict structure; an unforgettable presentation includes surprises, twists, turns, personality, and passion. Successful Science Pitches extend far beyond IMRAD. Instead of reciting dry statistics, tell the story behind the facts and figures to create special moments of suspense. A technically sound presentation with a storyline in which you bring in your passion is much more likely to inspire your audience than clinical-robotic content presented without emotion, in which only facts are conveyed.

You can use two easy-to-implement structures to create this arc of suspense in your Science Pitch: you will learn about Randy Olson's "Narrative Spectrum" (2015, pp. 115–123) and Emma Coats' "Pixar Pitch" (2011).

Finally, a convincing Science Pitch paves the way to funding. This opens new networking opportunities and enables you to advance your career successfully. In 2012, Dr. Mai Thi Nguyen-Kim came third at the Falling Walls Conference in Berlin— without applying the IMRAD structure! After her talk "Breaking the Wall of the Human Cell" (Falling Walls Lab 2012), a WDR employee invited her to a backstage day at the science program "Quarks" (Krapp and Schmermund 2019). This paved her way into television and science journalism. With her channel "MaiLab", Nguyen-Kim has reached over 1.4 million subscribers and more than 130 million views as a YouTuber; she stands out as a book author and later hosted "Quarks" herself for 3 years (Wikipedia 2024).

References

Coats, Emma 2011. Pixar Story Rules. Source: Price, David A. Stories from the Frontiers of Knowledge. Pixar Story Rules 2011. https://www.davida-price.com/pixar-story-rules. Accessed 13 Jan 2024.

Falling Walls Lab 2012. Dr. Mai Thi Nguyen-Kim. Breaking the Wall of the Human Cell. https://www.youtube.com/watch?v=5E2mIh4TA4k. Accessed 28 Oct 2024.

Krapp, Claudia and Schmermund, Kerstin 2019. Wissenschaftskonferenz – Pitch mir doch mal deine Forschung. Forschung & Lehre. https://www.forschung-und-lehre.de/detailview/pitch-mir-doch-mal-deine-forschung-2288. Accessed 15 Jan 2024.

Olson, Randy 2015. Houston, We Have a Narrative. Why Science Needs Story. University of Chicago Press, 256 pages (Paperback).

Wikipedia 2024. Mai Thi Nguyen-Kim. https://de.wikipedia.org/wiki/Mai_Thi_Nguyen-Kim#cite_note-24. Accessed 13 Jan 2024.

The First Elevator Pitch in History

Abstract

Learn how the engineer Elisha Graves Otis laid the foundation for modern urban development in 1854 with the first elevator pitch at the World's Fair in New York. Elevator pitches gained wider distribution, especially in sales, since the 1970s.

In the elevator pitch, you present your idea or product coherently, convincingly, and succinctly. Because your time is limited to the duration of an elevator ride, you only touch on a few facets of your topic. This will arouse curiosity in your audience and lay the foundation for further exciting information.

Do you know the story of the world's first elevator pitch? Let me take you on a journey back to 1854 when we are at the World's Fair in New York. At the time, taking an elevator was risky because it often crashed, and many people died.

At that time, engineer Elisha Graves Otis is developing a safe braking system to prevent such crashes, an ingenious idea with the prospect of safely transporting loads and saving lives. Otis rents the largest hall at the World's Fair and builds a highly visible elevator shaft. The crowd watches as he lets his assistant take him to the top. Otis looks briefly into the audience. His assistant smashes the rope with a sword. The platform hurtles downwards, and the people react in near panic. Barely a second later, the emergency brake is triggered, bringing Otis to a halt. He looks at the crowd and says: "All safe, gentlemen. All safe." (Otis 2023a, b) (Fig. 3.1).

This particular elevator pitch laid the foundation for modern urban development: the first skyscrapers with elevators were built after Otis founded the Otis Elevator Company. In 2023, it is the world's largest producer of elevator systems, with 71,000 employees and a turnover of 14.2 billion dollars (Otis 2024). Figure 3.2 summarizes these highlights regarding the first elevator pitch in history.

Fig. 3.1 Elisha Graves Otis delivers the first elevator pitch in history at the World's Fair in New York in 1854 (source: with kind permission of © Otis Elevator Company)

There is another version of the first elevator pitch: this was first created in the 1970s. Back then, young salespeople at a US company walked back and forth in front of the elevator of their office tower for some time until their CEO finally returned from his business trip. During the elevator ride of just under a minute to the office, they could excite their CEO about their new project idea. They made an appointment to present their project in detail and received funding approval (Skambraks 2012).

First Elevator Pitch

- **1854**: World's Fair in New York
- **Elisha Graves Otis** develops a safe braking system for elevators
- Visual effect: Simulated fall and words: "All safe, gentlemen, all save."
- Basis for **modern urban development**
- **Otis Elevator Company 2023**: The world's largest producer of elevator systems

Fig. 3.2 Key Takeaway: Elisha Graves Otis presents the first elevator pitch in 1854. It still has an impact today (source: author illustration)

References

Otis Elevator Company 2023a. Our First Passenger Elevator: This Day in Otis History. https://www.otis.com/en/us/news?cn=this-day-in-otis-history/. Accessed 13 Jan 2024.

Otis Elevator Company 2023b. Our history: A story of innovation and progress. https://www.otis.com/en/us/our-company/history. Accessed 09 Feb 2024. © Otis Elevator Company

Otis Elevator Company 2024. Fact Sheet. https://www.otis.com/documents/d/otis-2/otis-factsheet-final. Accessed 05 Oct 2024.

Skambraks, Joachim 2012. Sofortwissen kompakt: Elevator-Pitch. Heragon Verlag. 108 pages.

From Elevator Pitch to Science Pitch

4

Abstract

In an elevator pitch, you summarize essential information and capture your audience's interest. Start-ups and marketing and sales departments often present elevator pitches, usually to potential investors. The existing literature addresses corresponding business cases, emphasizing topic relevance, performance, and clear communication at various levels. It also points out typical practical mistakes.

The Science Pitch is a clear and convincing short presentation in which you concisely articulate your research project, reinforce it with a succinct take home message, and demonstrate your close personal connection to the topic. In short, a Science Pitch is precise and understandable. It is also persuasive and personal.

The Science Pitch targets experts and academics who want or need to get to the point of introducing their project. You develop your short presentation based on the ESPRIT model introduced here in a Science Pitch Canvas. By presenting your project and demonstrating your expertise, you inspire your audience, succeed in Science Pitch competitions, and secure funding for your research project.

Supplementary Information The online version contains supplementary material available at https://doi.org/10.1007/978-3-658-44844-8_4. The videos can be accessed individually by clicking the DOI link in the accompanying figure caption or by scanning this link with the SN More Media App.

13

S. Wagner, *Science Pitch*,
https://doi.org/10.1007/978-3-658-44844-8_4

The *Elevator Pitch* is not about immediately recounting all the details and facets. Instead, get to the heart of the essential information to raise your conversation partner's or audience's curiosity. You can go into more detail if they want to learn more about your idea.

Nowadays, entrepreneurs, founders, and applicants have long used elevator pitches outside of an elevator. They are suitable as a kick-off to a more extended collaboration or for positioning yourself at networking events. Job interviews and the start of a telephone call are also suitable for using elevator pitches.

Pitches are flexibly used in marketing and sales, depending on the occasion: from a slogan in one or two short sentences to a pitch deck in which the founders deliver a 5-min presentation. This is followed by a further 5 min, during which they answer questions from potential investors. They are invited to a more in-depth discussion of the content details if they gain interest.

Most elevator pitches relate to business cases, especially for start-ups and founders pitching to potential investors or sponsors. Therefore, it is unsurprising that practically all books on elevator pitches have been written precisely for such settings.

For example, US National Elevator Pitch Champion Chris Westfall (2012) presents scenarios for TV shows and investor pitches, job interviews, and conversations about salary increases. It is about "seven steps to clarity" (Westfall 2012, Part I, p. 18), including authenticity, relevance, inspiration, and tact. In contrast, Skambraks (2012) focuses on conversations with investors and customers. The references to networking events and opening personal conversations are particularly interesting here. However, he essentially limits himself to very short elevator pitches of 20–30s.

More recent literature also refers to elevator pitches in the entrepreneurial and start-up environment. Becket (2019), for example, summarizes the most important elements of pitches in a specially developed Pitch Canvas©. The book is aimed at start-ups and innovation teams in companies. Funken and Altenschmidt (2021) also address entrepreneurs and start-ups pitching to potential investors. In addition to common aspects such as relevance, slides, and performance, they also address the solidity of the company, the remuneration model, and the technical implementation, which are essential for elevator pitches. Grytzmann and Lexa (2021) take this one step further: they discuss several concrete examples of elevator pitches from practice in detail. In doing so, they uncover typical practical mistakes and provide concrete tips for better elevator pitches. This is about business pitches in front of customers, business partners, and investors. With their "Corporate Pitch", Gundlach and Fricke (2022, pp. 16–17) specifically address the business sector with companies, start-ups, and freelancers. This involves "contracts put out to tender and evaluated by a corporate committee". They differentiate between the "horizontal pitch" in a competitive situation and the "vertical pitch", in which a "joint path into the future" is shown.

It quickly becomes apparent that none of these books deal with elevator pitches for science presentations. The basics for successful pitches, such as the topic's relevance, pitch structure, and aspects of a convincing presentation are essentially the same. But how can you summarize often complex content from research projects so that the interest across disciplines is aroused and the scientific expertise quickly becomes clear? What can experts, especially academics, from students to doctoral candidates, research assistants to professors, do to inspire their audience?

There are at least three books that deal more particularly with science presentations. One classic is Olson's book about storytelling in science (2015). We specifically address his "Narrative Spectrum" presented there (Olson 2015, pp. 115–123), which is now quite well-known and easy to implement as a valuable component of the Science Pitch. It is used in many storytelling and writing seminars for doctoral students, as participants in my training courses repeatedly report. Very sound information on presentations in science can be found in Browning (2021). However, Browning mainly refers to longer presentations and only briefly touches on pitches for the discussions that often follow presentations. Finally, Hey's (2023) latest edition on science presentations provides comprehensive information on preparing and delivering persuasive speeches.

The video highlights that in contrast to the cited literature, this book deals explicitly with *Science Pitches* (Fig. 4.1). It thus sets a unique focus for times in which scientists and other experts must be able to break down comprehensive projects and experiences to the essentials in the shortest possible time and at the same time attract, convince, and inspire their audience. Elevator pitches have long since arrived in the academic world and are becoming more and more established. A short talk presented here is the Science Pitch. Common formats include 3-min poster presentations at science conferences and competitions such as "FameLab" (FameLab Switzerland 2023) and the "Falling Walls Lab" (Falling Walls 2011–2024), as well as numerous other events with Science Pitches for spin-offs from research projects.

A short and clear take home message is essential for a convincing Science Pitch. You also respond to important questions such as "Why are you doing research?", "What motivates you?" and "Who will benefit from this project?". Unlike longer presentations or lectures, the Science Pitch is about the essence of the information: you emphasize the specific added value and benefits of your project and its practical relevance. Many experts fail to establish a personal connection to their data, facts, and figures!

> The Science Pitch is an easily understandable and convincing short speech in which you get to the heart of your research project in the given timeframe, underpin it with a concise take home message, and demonstrate your close personal connection to the topic. In short, a Science Pitch is precise, comprehensible, convincing, and personal.

But how can you present your project precisely and be understood immediately? How do you convince your audience? How do you share your enthusiasm for your topic? That is what this book is all about.

The Science Pitch is aimed at experts, especially academics, from students to doctoral candidates and research assistants to professors. In the academic field, people like to work with models. Models provide an initial orientation and can be improved and adapted continuously. This is how you develop your Science Pitch

Fig. 4.1 From elevator pitch to Science Pitch (source: author illustration)
(▶ https://doi.org/10.1007/000-cyz)

by applying a newly introduced model: the ESPRIT model. With
ESPRIT, you can create your Science Pitch Canvas, develop the
content of your short presentation, and adapt it to the prevailing
situation. You focus on the most essential and exciting content for
your audience. This will enable you to skillfully combine your
scientific expertise with special moments of innovative research.
You will learn about typical pitfalls and develop high-class short
presentations in which you bring your expertise with power to the
point. You will inspire your audience, impress in Science Pitch
competitions, and be awarded funding for your research project
(Fig. 4.2).

Science Pitch

- ⸕ **Science Pitch**: Short presentations, posters, project proposals, networking events, contest
- ⸕ **Goals**: Arouse curiosity and get appointment
- ⸕ **Why and what for?** Motivation and benefits
- ⸕ **Essence** of information, **practical relevance**
- ⸕ **Story** behind data, facts, and figures
- ⸕ **Personal connection** to the project topic
- ⸕ **Focus** on the lifeworld of **audience**

Fig. 4.2 Key Takeaway: Science Pitches are short presentations with a succinct take home message (source: author illustration)

References

Beckett, David 2019. So gewinnt man jeden Pitch. Die optimale Vorbereitung, um Kunden, Kollegen und Investoren zu überzeugen – mit vielen Praxistools und Fallbeispielen. 224 pages, Redline Verlag.

Browning, Jo Filshie 2021. Scientifically Speaking. 168 pages, Practical Inspiration Publishing.

Falling Walls 2011–2024. https://falling-walls.com/lab/. Accessed 13 Jan 2024.

FameLab Switzerland 2023. https://www.famelab.ch/. Accessed 13 Jan 2024.

Funken, Irmengard and Altenschmidt, Karsten 2021. Perfekt im Pitch. Kunden begeistern, Investoren überzeugen. 126 pages, Haufe-Lexware GmbH & Co. KG.

Grytzmann, Oliver and Lexa, Carsten 2021. So gewinnen Gründer ihre Pitches. Kunden, Geschäftspartner & Investoren mit gelungenen Präsentationen überzeugen. 125 pages, Springer Fachmedien Wiesbaden GmbH.

Gundlach, Nico and Fricke, Lukas 2022. Die Kunst, Menschen zu begeistern. Wie man Aufträge gewinnt mit dem perfekten Corporate Pitch. 240 pages, Redline Verlag.

Hey, Barbara 2023. Präsentieren in Wissenschaft und Forschung. In Präsenz und virtuell. 209 pages, Springer Gabler Wiesbaden, 3rd Edition.

Olson, Randy 2015. Houston, We Have a Narrative. Why Science Needs Story. 256 pages, University Press.

Skambraks, Joachim 2012. Sofortwissen kompakt: Elevator-Pitch. 108 pages, Heragon Verlag.

Westfall, Chris 2012. The new Elevator Pitch. 214 pages, Marie Street Press.

Networking Events: From Clear Targets to Strong Connections

5

Abstract

Networking events are particularly beneficial once you have a clear target. By preparing specifically, you can find suitable conversation partners more quickly and provide even more added value. Identify thematic connections and consider appropriate conversation starters that will lead from general small talk to more personal Big Talk.

Adapt your conversations so that you are well understood. Use simple and universally comprehensible language for the interested public, more advanced language for your professional audience, and a combination of both for a mixed audience. Whether it is a Science Slam, a PechaKucha talk, your PhD thesis defense, a research proposal, or a job interview, your networking conversations and your Science Pitch should be clear in content but never literally scripted in preparation.

Networking events take you further the better you know your goal. By preparing specifically, you can offer your conversation partners even greater added value (Fig. 5.1).

© The Author(s), under exclusive license to Springer Fachmedien
Wiesbaden GmbH, part of Springer Nature 2024
S. Wagner, *Science Pitch*,
https://doi.org/10.1007/978-3-658-44844-8_5

Answer the following questions in advance:

- What exactly does this event offer me?
- What is my personal goal, and how do I achieve it?
- What makes me interesting as a personality or as an expert on a topic?
- Which topics will I specifically inform myself about beforehand?
- Why do other people attend the event?
- Whom do I want to meet for the first time? For what purpose?
- Do I want to meet certain people? What do I want to talk to them about? What specific added value can I offer them?

It is worth your time to answer these questions. Whether you are pursuing your PhD, looking for collaboration partners, or aiming for a new position, the better you know your personal goals and the intentions of your conversation partners, the more effectively you can align both.

Make notes on what you want to talk about. This can vary significantly in individual cases. What do you want to discuss with your target audience? Conversely, ask yourself which of these aspects fit your presentation. This way, you can also weave in comparative studies or appropriate references.

Consider which topics you can start a conversation on. What personal connection do you create with your conversation partners? This is not about the weather.

More suitable are classic small talk questions like:

- How did you like the presentation of ...? What are you taking away for yourself?
- Are you also delivering a talk? On what topic?
- Which project are you currently working on? What excites you about it?

If your gut feeling fits and you feel comfortable, you can also ask Big Talk questions. These are particularly open and thought-provoking, such as:

- What is your opinion on the topic of ...?
- What do you do differently than others—and why?
- What has brought you the most progress (in your career/life)—and why?
- What can you do today that you could not do a year ago—and why?

With questions like these, you can skip the often rather superficial small talk. The idea of Big Talk was developed by entrepreneur and journalist Kalina Silverman (2024). Big Talk is much more personal and direct than small talk. Used at the right moment, it can deepen your interpersonal relationships and lift them to a higher level. In this way, you learn much more about the people around you. Big Talk is not a one-way street: you should also be open to sharing personal details about yourself.

You can also introduce your topic or project to such discussions in the form of a Science Pitch. Researching potential discussion partners in advance makes sense, as this will help you find thematic links more quickly.

Whether it is a discussion or a presentation, if you know your audience and their motivation for attending your talk, you can emphasize specific content and facets of your topic. In front of a specialist audience, scatter in individual scientific terms. When addressing the wider public, use simple language everyone can understand.

Does your audience consist of both groups? In this case, you can start by introducing your topic in broader terms. Then, present one or two facets of it in more depth. In a Science Slam or a PechaKucha talk, you can present your topic in a particularly relaxed, colloquial, and entertaining way. If you are defending your PhD thesis, immerse yourself much more deeply in the subject. When you present your research proposal, make additional references to neighboring disciplines and practice. Find out who will review and assess your proposal to ensure you can introduce thematic links.

In the job interview, on the other hand, you present the most exciting aspects of your project for your future employer: what practical experience do you bring to the table? How can you use this for new projects? How will your expertise contribute to your new job?

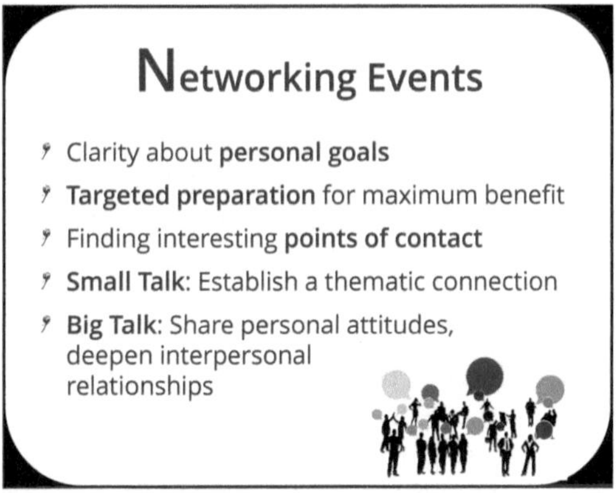

Networking Events

- Clarity about **personal goals**
- **Targeted preparation** for maximum benefit
- Finding interesting **points of contact**
- **Small Talk**: Establish a thematic connection
- **Big Talk**: Share personal attitudes, deepen interpersonal relationships

Fig. 5.1 Key Takeaway: Prepare yourself specifically for networking events (source: author illustration)

Reference

Silverman, Kalina 2024. Big Talk. https://www.makebigtalk.com/. Accessed 13 Jan 2024.

Get to Know Your Audience in Advance

6

Abstract

Through targeted research, you will better understand your target audience. This lets you craft your Science Pitch precisely and find surprising links with special added value. Inform yourself about current developments and news in your professional field. Find out what drives your conversation partners and your network. This way, you can work out specific questions and create special added value for your contacts with your expertise.

Your preliminary research will help you to get to know your target audience better and thus prepare your Science Pitch precisely. Is there a program for the event with a list of speakers and participants? Perfect! This will give you an initial overview of the event's other topics. You can establish links to these topics or differentiate yourself in content based on this information.

You can research your most exciting contacts on the internet. Their online profiles on social media platforms such as LinkedIn, ResearchGate, Instagram, Facebook, TikTok, and YouTube will provide some valuable information. You can find further information on personal websites, on the websites of the respective employers, or in press reports.

Most interesting is information such as …

- *Personal presentation.* Career background, professional expertise, and specific qualifications.
- *Mutual acquaintances* and second-degree contacts.
- *Interviews.* On what topics do they comment? What is their language style and jargon? How do they dress? In what environment do they spend time?
- *Projects and publications.* What do they write about in publications? How do they express themselves? What do they reveal about themselves?
- What *personal preferences and hobbies* do they share online?
- What does the current *market and industry development* in the according sector look like?

This way, you gain an initial overview of their profiles and networks. You can strategically incorporate this information into your conversations. By addressing similar professional experiences and hobbies, you quickly establish rapport. Most importantly, you gain new ideas and references for the content of your Science Pitch. Adapt information flexibly for an ideal personal match.

6.1   How to Provide Tangible Added Value

Network meetings are not a one-way street. Although your Science Pitch is a stand-alone monolog, it creates space for personal discussions. Clarify for yourself the value you can add to your conversation partners. Otherwise, their interest in time spent together may remain limited.

Do you know in advance who is attending your presentation? Do you know what your audience is interested in (Fig. 6.1)? Your thematically established peer group expects more in-depth content than an interested audience that lacks detailed specialist knowledge. The same applies to your language style: predominantly easy to understand for a mixed audience, more scientific terms for experts in your field.

To add specific value to your contacts, answer the following questions:

- Do I know the participants of the network? Do I know about potential collaboration partners and their personal stories?
- Why and for whom am I an interesting person?
- Why are my project, my product, or my service interesting? How does it make a positive difference compared to others?
- What added value do I provide as a person and personality to my environment? What good am I doing? Do I incorporate this into my conversations?
- What professional expertise do I bring with added value?
- What life experience do I bring to others?
- What questions do others ask again and again?
- What should my audience think, feel, and do after my Science Pitch?

A goal-oriented approach at networking meetings will help you become better known in your field. You will gain practical experience in presenting yourself and your current project interestingly, enabling you to further develop your Science Pitch in a targeted manner.

6.2   Real-Life Example: Networking at Science Conferences

Many participants in symposia are interested in the latest developments and news from their professional field. They attend conferences to exchange ideas and maintain and expand their network. Some are specifically looking for potential collaborations. Others look around for new career prospects and approach potential employers or employees. The vast majority present the status of their project. Those with booths in the exhibition area want to promote and sell their products and services.

What is your goal when you attend conferences? You can prepare yourself specifically if you don't want to be driven by chance:

- *Study the program* and select the most exciting topics in advance. Why are these topics of particular interest to you?
- *Visit the profiles of the speakers and experts* to learn more about their specialist areas and personal interests.
- *Prepare your questions in advance.* Think about the input you can contribute to discussions. What interests other people about you?
- *Make contact well before* the conference and arrange meetings for personal discussions.

Always remember you have something valuable to share!

6.3   Real-Life Example: Confident at Eye Level in Job Interviews

During a job coaching session, I researched with the applicant how the company's HR manager presents herself online. We were lucky: in addition to a photo on the website and a short CV, we also found a video of her dressed in business casual style and answering questions in an interview. This allowed us to gain a first impression and brief insight into the personality and behavior of the HR manager's personality and behavior.

I once applied for a position as a research assistant in Australia. For the assessment center, I wanted to know who would be sitting next to me in the interview. Thanks to my research, I correctly estimated five out of six institute employees. I was able to investigate their research topics, publications, and other interests; the interview went well because I was able to specifically address the professional preferences of the respective scientists and talk about them in the subsequent discussion, also making links to my topics.

When applying to companies, read the annual reports and press releases from the recent past and find out about current developments in the relevant sector. You should also re-read the job advertisement: what tasks are described there? Which qualifications are essential? These are valuable pointers to possible content for your short self-presentation when asked to "Tell us about yourself ...".

Your Target Audience

- ⸕ **Research online in advance**: What you learn about people in your target group
- ⸕ **Program**: Who and what is interesting?
- ⸕ **Topic research**: Latest developments and discussions
- ⸕ **Prepare specific questions**
- ⸕ What specific **added value** can you pass on?

Fig. 6.1 Key Takeaway: Learn the most about your target audience (source: author illustration)

Typical Pitfalls in Science Pitches

7

Abstract

Fact-based presentations are especially important in the science community. However, what often lacks is a personal connection to the topic. For a convincing Science Pitch, focus on particularly interesting content. Share special moments in your research and your vision, mission, and values.

Adopt your audience's perspective and find direct content connections to neighboring disciplines. Get to the essence of your topic without exceeding time limits.

Analyze your manner of speaking: Direct addresses like "you" and "we" sound more personal than a distant "one". Short sentences alternating with longer ones are more engaging than long, complex sentences. Translate facts into memorable stories and anecdotes. Reduce text on slides to a minimum and instead use images and graphics that are easy to grasp.

Gain new insights through personal conversations well before your presentation. With a well-prepared plan B, you can modify your Science Pitch to unforeseen situations. Arrive early at the venue and familiarize yourself with the technical setup—including online settings. Prepare for audience questions.

© The Author(s), under exclusive license to Springer Fachmedien Wiesbaden GmbH, part of Springer Nature 2024
S. Wagner, *Science Pitch*,
https://doi.org/10.1007/978-3-658-44844-8_7

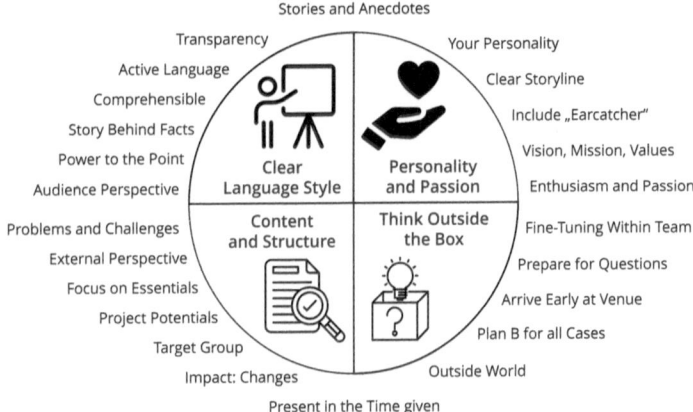

Fig. 7.1 The 24 typical pitfalls in Science Pitches (source: author illustration)

Do you deliver short talks? Do you listen to other presentations? Then you probably know the four most common pitfalls that undermine a well-conceived presentation: (1) The presentation is impersonal and lacks passion. (2) The structure of the speech remains ambiguous. (3) The speaker does not think outside the box. (4) He or she makes insufficient use of the linguistic and visual potential.

I have encountered some of the following 24 typical examples (Fig. 7.1). Here, I show how you can do better than most speakers.

7.1   Inject Personality and Passion

One of the most frequent pitfalls in any presentation, especially in science and business, is the speaker's failure to connect personally to the project presented.

- *Show Your Personality.* Many limit themselves to presenting their projects with facts, figures, and technical content. Few show a personal connection to the topic; they avoid revealing their enthusiasm. But how can we inspire our audience if we remain factual and neutral? The solution: share anecdotes and personal stories directly connected to your topic! Your audience should feel that you identify with your project and are motivated.

- *Share Your Enthusiasm, Your Excitement, and Your Passion.* Successful presentations need emotions! Why do you think your project is so cool and so compelling? Can you share the special moments of your research in the Science Pitch? What excited you as a child? Do you remember a situation in which you gained a groundbreaking insight or a discovery? These are precisely the moments when your eyes light up! They also create a bond with your audience.
- *Demonstrate Your Motivation.* What is your vision for the project? Do you have a clear mission? What values do you represent? Few people address these aspects in the presentation. We will cover them in detail in the "storyline" section (see Chap. 12).
- *You Need a Clear Storyline.* Stories and anecdotes require an entertaining arc of suspense. The most straightforward sequence is situation—problem—solution—insight—outlook. Olson (2015) uses the simple structure "and—but—therefore" for this. The AND connects individual findings and provides the all-important context between these results. The BUT creates suspense: something new and surprising comes into play here, challenging state-of-the-art knowledge. This also changes the direction of the narrative. "THEREFORE" addresses the consequences, so you reach a new level of knowledge with your conclusion. The result is something new that moves your audience forward!
- *Introduce "Earcatchers" in Your Presentation.* Richter and Münzner (2020, pp. 103–104) point out news anchors. They start their reports with an "earcatcher" to immediately gain the attention of their viewers. Our audience is particularly attentive when we "... present an incredible figure, news not thought possible, a provocation, a beautiful picture, a peculiar comparison, astonishing facts". Here are two exemplary earcatchers for a presentation: "We have discovered what researchers have been

looking for for decades ..." or: "Our results raise completely new questions about extraterrestrial intelligence ...".

7.2   Address Only the Most Important Content and Ensure Clear Structure

Many presenters want to cover as much content as possible yet neglect to focus on the most exciting essentials. Some do not even cover future perspectives.

- *Directly Address Your Target Audience.* Take the perspective of your audience. Address their questions or touch on them instead of sharing your perspective only. In job interviews, most candidates miss out on putting themselves in the employer's position. Your Science Pitch is more convincing if you can take both perspectives.
- *Broaden Your Perspective.* Do you consider groups outside your field? Avoid the tunnel vision and answer these questions: are there intersections between your project and neighboring disciplines? What relevance does it have for social, political, economic, and ecological concerns? Bayley and Phipps (2019, p. 10) define "Research Impact" as "provable change (benefit) of research in the 'real world'". The Director of Research and Development at the University of Lincoln explains that project findings ideally "go beyond the academic walls" (Bayley 2018). The best example is the health sector: researchers have developed new COVID-19 vaccines and brought them to market. Do you communicate your results excellently and in a manner that is generally understandable to a public audience? Then, you have the chance for awards like the Communicator Prize of the German Research Foundation (DFG), which stands "for successful dialog between science and society" (Stifterverband 2024).
- *Address Problems and Challenges.* What problems did you face during the project? What challenges have you mastered already, and which ones may still arise? Discuss it briefly and concisely.

- *Discuss the Impact of Your Project.* Go into what changes are due to your project. Have you developed something new? How can you apply your findings in practice? Can you scale your innovation? You can also refer to a patent you have developed yourself. There are further possibilities for knowledge transfer and societal benefit. This reaches far beyond mere data, facts, and figures. The latter already take up too large a part of presentations anyway.
- *Talk About the Potentials Your Project Opens Up.* What new opportunities arise from your findings? Have you developed a product prototype and put it into serial use? Address this as precisely and plausibly as possible. Again, this sets you apart from the broad mass that does not even recognize the potential for follow-up projects based on their project.
- *Focus on the Essentials.* Resist the urge to cram too much content into your limited presentation time. In the Science Pitch, you narrow down your content to key aspects. This increases your audience's curiosity and prompts targeted questions. You can address the details of your project after your Science Pitch. This way, you gain focus and avoid rambling on generalities. Less is more!
- *Use Your Allocated Time Appropriately.* If you have 3 min for your Science Pitch, do not finish after 1 or 2 min. Otherwise, you unnecessarily give away time for such valuable information about your project and expertise. Also, do not overrun: your audience is often more impatient and relentless than you imagine. Be so well prepared through repeated rehearsal that you close your Science Pitch after 2:50 min, at most after precisely 3 min, without rushing at the end.

7.3 Language Style and How to Present Convincingly

Another major pitfall while presenting research or business projects is failing to synchronize language style with your target audience. Instead of a mere list of facts, they fail to translate their results into compelling stories.

- *Spoken Language Sounds Personal.* Spoken language is more open and direct than written language, which is comparatively passive. Write down your pitch as if you were speaking directly to a good friend. Test your pitches by speaking freely: record yourself using a voice recorder, such as the app built into your smartphone. You can transcribe the text or write it out. Alternatively, open MS Word and record the text directly using the "Dictate" function.

- *Speak in Simple, Understandable Terms.* If you can explain complex content in a scientifically sound and, at the same time, generally understandable way, you demonstrate your communication skills. Replace supposedly great-sounding buzzwords or hackneyed phrases with clear and meaningful expressions. If your target audience knows the topic, you can use technical terms discreetly. With a directly formulated manner of speaking, you ensure that you connect with your audience much more quickly than if you talk indirectly. This makes you sound much more personal, active, and at the same time more reliable than with indirect, distancing language: "We ..." instead of "One ...". "Here we have ..." instead of "Here one has ...", "We were able to investigate this reliably by ..." instead of "The reliability of the investigation is guaranteed ...". Avoid empty phrases and trivial sentences such as "The present work serves to investigate the causes of ..." or "Our results imply for practice ...". Instead, use metaphors and figurative language. In this way, you sharpen the content, which you can combine with technical language if necessary.

- *Always Be Transparent in What You Say.* In pitch competitions, you must elaborate on important content. Communicate the goal and status of your project clearly and transparently. At least briefly touch on the information that is essential for your audience. Avoid nested sentences, thereby presenting content in an unnecessarily complex way. Olson (2015) pointed out that narrative structures such as „and—and—and" and "despite—however—yet" are tedious or overly complex. A structure like "and—but—therefore" suits an entertaining presentation much better. Use relatively short and only occasionally longer sentences. This will keep your audience engaged and listening actively instead of dropping out of your presentation entirely.

- *Tell the Story Behind the Data, Facts, and Figures.* There is more to it than the popular term "storytelling": your information is far more memorable if you translate abstract content into pictorial stories and emotions.

Example: do you prefer to see a graph or table with numerical values? Or do you prefer a visual animation to a statement like: "The rain infiltrating the soil dissolves and displaces the carbonates. Over time, they are carried out into rivers and oceans."?

If individual aspects are discussed controversially, you can provide vivid pro-and-con examples before sharing your assessment.

- *Combine Factual and Rational Information with Stories and Anecdotes.* Do it like David McCandless and Hans Rosling! Both have been on stage as TED speakers and have inspired their audiences. At the end of his presentation on the beauty of data visualization, McCandless (2010) talks about Eyjafjallajökull, the volcano that erupted lava in Iceland in 2010. He compares the amount of carbon dioxide (CO_2) emitted by this event with its savings from air travel, which was inactive at the time. The amounts were similar. There is no better way to summarize this information than in his words, "We had the first carbon-neutral volcano." The audience reacts with appropriate enthusiasm before McCandless concludes, in keeping with the title of his talk, "And that ... is beautiful." On the other hand, Hans Rosling (2006) takes his time to explain a complex graphic in detail. He provides his audience with comprehensibly explained basic information. Only when the basis is established does Rosling animate the graphic and speak enthusiastically, acting simultaneously as a conductor. I have analyzed this TED Talk in a blog article (Wagner 2020b). Both presentations thrive on the speakers analyzing data sets and translating them visually. Above all, they put developments into a larger context. Both narratives reveal new contexts in an appealing visual design. By changing how we view the world, they stand out as distinct, high-level presentations among numerous other talks.

- **Present with Power and to the Point.** Do you present at an average level like the masses? Or do you want to stand out positively so your audience will remember you? Then, reduce the text on your slides to an absolute minimum and use the full potential of visual information! Depending on the event, you may be allowed just three slides in your Science Pitch. Choose carefully which information is the most interesting, impressive, and important, and show only this on your slides. Can your audience grasp the content of your slides in 3 s or less? That rules out complex presentations from the start. David McCandless and Hans Rosling are setting new standards here as well. I go into this in detail in my blog (Wagner 2022).

7.4   Think Outside the Box

The speaker should prepare well before the presentation and familiarize themselves with the technical aspects to remain flexible during last-minute changes.

- **Get in Touch with the Outside World and Your Target Group in Advance.** Some people work on their presentation alone in isolation. They miss the unique chance to talk to the outside world long before their presentation. Their message will rarely reach their audience. Instead, approach your target audience early on: discuss open questions and problems, ask for their perspective, and incorporate some of their ideas and needs. This can lead to completely new solution approaches. In the follow-up meeting, you can introduce other exciting aspects of your project as needed. These conversations allow you to test whether your vision or the goal of your project corresponds to your target group. Possibly, new focal points for your Science

Pitch will emerge. Take it one step further: answer the question of how your project relates to neighboring disciplines, the market, or user behavior. Which new opportunities emerge this way?

- *Arrive at Your Presentation Venue More Than on Time.* What time will you arrive at your presentation venue? I occasionally see presentations where speakers rush to the event only at the last second before their presentation begins. They must get familiar with the room and prepare the technical setup first. So, your audience would have to wait unnecessarily: these are not good conditions for a convincing presentation. Often, there are still conversations with the host, which further shortens your preparation time. So, plan for sufficient time in advance. If you are presenting online, ensure your technology works smoothly and your setting is optimal well in advance. Optimal sound quality and the fitting image detail are particularly important. The program you use should be kept up to date; updates should be installed well in advance. You can get detailed information in my blog (Wagner 2020a).

- *Coordinate with Your Team if Multiple People Are Presenting.* Well-coordinated content and the right dramaturgy count. Think carefully about the order in which you present your pitches. Make sure to create well-thought-out suspense for your audience with smooth transitions without unnecessary repetitions. The TV show "Dragons' Den" provides many good examples.

- *Prepare for Audience Questions.* Presentations, especially pitches are often followed by a discussion of the content or at least a question-and-answer session. Are you prepared for critical questions? Do you know the potential pitfalls in discussions? If surprising objections or attacks completely throw you off your game, this can derail many a project you thought was already securely funded. This will not happen if you prepare yourself well. You can avoid some of the expected questions by answering them during your pitch already. These include important methodological aspects and questions that force you to think outside the box (Fig. 7.2). What surprising questions do experts in your field ask you? What questions will you get from people who are unfamiliar with your topic?

- ***Know Your Plan B, Just in Case.*** Familiarize yourself with the technology beforehand: check the microphone and the sound and be prepared to present without slides and videos. Online, clear sound and a stable internet and video connection under optimal lighting conditions are essential. Can you spontaneously bring forward your presentation, appear later than planned, or shorten the presentation time? All this is part of your professional preparation. This way, you can always deal with unexpected incidents confidently and skillfully. Despite intensive preparation and arrangements, I had to adjust my time budget again and again spontaneously, speak at a completely different time than planned, or present without slides at all. Thanks to years of experience and because I had a backup plan, I could switch quickly.

Regardless of your pitfalls, adopt the mindset of Seth Godin, U.S. entrepreneur, author, and marketing expert. Godin (2012) says, "The best elevator pitch is true, stunning, brief and it leaves the listener eager (no, desperate) to hear the rest of it. It's not a practiced, polished turd of prose that pleases everyone on the board and your marketing team, it's a little fractal of the entire story, something real."

Avoiding Pitfalls

- ⸙ Share **personal passion** in presentations
- ⸙ Only address the **most important content**
- ⸙ Ensure a **clear structure**
- ⸙ Convince through a **clear verbal style**
- ⸙ **Present with power to the point**
- ⸙ **Thinking outside the box**
  can dramatically increase
  the quality of your presentation

Fig. 7.2 Key Takeaway: How to avoid pitfalls in your Science Pitch (source: author illustration)

References

Bayley, Julie 2018. Sausages, unicorns and strip clubs. Or Impact: the challenge of connection. https://juliebayley.blog/. Accessed 13 Jan 2024.

Bayley, Julie and Phipps, David 2019 (revised 2022). Extending the concept of research impact literacy: levels of literacy, institutional role and ethical considerations [version 2; peer review: 2 approved] Emerald Open Research 2019, 1:14. https://doi.org/10.35241/emeraldopenres.13140.2, First published: 2019, June 07, 1:14 https://doi.org/10.12688/emeraldopenres.13140.1. Accessed 13 Jan 2024.

Godin, Seth 2012. No one ever bought anything on an elevator. https://seths.blog/2012/10/no-one-ever-bought-anything-on-an-elevator/. Accessed 13 Jan 2024.

McCandless, David 2010. The beauty of data visualization. TEDGlobal 2010. https://www.ted.com/talks/david_mccandless_the_beauty_of_data_visualization?referrer=playlist-how_to_make_a_great_presentation. Accessed 13 Jan 2024.

Olson, Randy 2015. Houston, We Have a Narrative. Why Science Needs Story. University of Chicago Press, 256 pages (Paperback).

Richter, Kay-Sölve and Münzner, Richard 2020. Viel mehr als nur Körpersprache – Executive Presence. Wie Sie als Führungskraft überzeugend auftreten, wenn es darauf ankommt. GABAL Verlag GmbH Offenbach, 240 pages.

Rosling, Hans 2006. The best stats you've ever seen. TED2006. https://www.ted.com/talks/hans_rosling_the_best_stats_you_ve_ever_seen. Accessed 13 Jan 2024.

Stifterverband 2024. Communicator-Preis. https://www.stifterverband.org/communicator-preis. Accessed 13 Jan 2024.

Wagner, Stephen 2020a. Video and sound check for online presentations. https://redelandschaften.de/en/video-and-sound-check-for-online-presentations/. Accessed 03 Feb 2024.

Wagner, Stephen 2020b. Stage your presentation and attract your audience. https://redelandschaften.de/en/attract-your-audience/. Accessed 03 Feb 2024.

Wagner, Stephen 2022. Use Power to get to the Point with your Presentation. https://redelandschaften.de/en/use-power-to-get-to-the-point-with-your-presentation/. Accessed 03 Feb 2024.

Integrate Artificial Intelligence (AI) Tools Like ChatGPT

Abstract

Artificial intelligence (AI) is currently established in our working lives. It can lead to a faster and more productive way of working than ever before. From a scientific perspective, you can thus formulate new hypotheses, design experiments, systematically collect and analyze data, prepare publications more efficiently than before, and prepare presentations like your Science Pitch.

The quality of AI-generated responses improves the more precisely you phrase your prompts. Ensure you use an actively formulated, preferably spoken text that matches your speaking style. This way, you capture attention and interest and demonstrate that you can skillfully sharpen scientific matters in content.

A practical example will teach you how you can design parts of your Science Pitch through ChatGPT.

The application of artificial intelligence (AI) has increased rapidly since the publication of ChatGPT on November 30, 2022 and is becoming more and more established in our everyday lives. AI is undoubtedly also transforming scientific work (Frueh 2023). It provides new possibilities, such as comprehensive simulations and personalized education. Below, we discuss how you can use

AI in academia and put it into practice—particularly for creating your presentations and Science Pitches.

8.1   Artificial Intelligence (AI) in Science

By using AI in the workplace, we can work even faster and more productively than ever before. The targeted use of AI for creative and analytical tasks generates a "gain in performance" (Lang and Hütter 2024, p. 39). Business psychologist Prof. Ulrich Lenz emphasizes that it is about the ability to "use AI to add value" (Lang and Hütter 2023a, p. 37). This applies to knowledge-intensive work and communication (McKinsey 2023, cited in Lang and Hütter 2023a). According to futurologist Prof. Pero Micic, AI language models now have "comprehensive knowledge" with a "depth of information" that can no longer be replicated even by the best experts (Lang and Hütter 2023a, p. 36).

AI can be used in science to hypothesize, design experiments, and systematically collect data. At the same time, however, the unbiased use of distorted, non-representative, and non-transparent data sets is critically viewed (Erduran 2023). Science must, therefore, always remain comprehensible, transparent, and ultimately a reliable basis for all communication, even with AI.

8.2   Artificial Intelligence (AI) in Practice

In practice, you can use AI to prepare new publications more efficiently than before. One option is targeted and faster literature searches with programs such as Perplexity (perplexity.ai), which work much more precisely than ChatGPT (as of February 2024). ChatGPT assists with the targeted preparation of presentations and Science Pitches. The program can create outlines and concepts for publications, project design, and courses using prompts that assign a task or query to the AI system. You can use these to develop your ideas and then refine and specify them individually.

The quality of the AI's answers increases the more precisely you phrase your prompt. The latter can, therefore, sometimes be quite extensive. "The requirement for precision at the factual level also applies to the desired output" (Lang and Hütter 2023b, p. 40).

The draft prompt could be something like this:

> Please write the following text in a [name style] style that is understandable and appealing to a [name target audience].

The style you could use here could be formal, informal, scientific, technical, humorous, empathetic, critical, commanding, or child-oriented. The target audience could be scientists, specialists, engineers, the public, an international audience, or children of a certain age.

Two examples of prompts would be:

Please write the following text in an informal style so that it is technically sound and immediately understandable for a scientific audience.
Please write the following text in an entertaining style to be understandable to the general public.

Now, insert the corresponding text before sending your prompt to ChatGPT.

You could also rewrite your text to prepare your research proposal. Or you could prepare your business elevator pitch for potential investors. Alternative, you could use it for a Science Slam, rewrite it in Goethe's version, or write it as a poem.

Of course, the AI can rewrite entire texts or delete unnecessary filler words and repetitions of content without replacement. The AI can then capture the essence of these and essential ideas in short, original, comprehensible sentences. However, you must first let your thoughts and ideas run free (Lang and Hütter 2023c).

As a scientist and expert, you must express yourself verbally so that an AI implements the prompts as accurately as possible for your purposes. It is, therefore, also about verbal precision as part of trainings for presenting elevator pitches and Science Pitches. The input or prompts to AI, such as ChatGPT, should also correspond to your natural speaking style. The AI takes care of the rest and precisely suggests phrases precisely that fit your specific occasion and target audience.

In practice, you can also spontaneously present your Science Pitch and record it with the ChatGPT app to immediately receive the complete text in writing. You then ask ChatGPT to eliminate all filler words and make any further stylistic improvements to the text. With this procedure, you will quickly have a spoken and thus actively formulated, true-to-life text instead of a text that is often passively formulated and typed by hand (Lang and Hütter 2023c).

In a self-experiment, I used the ChatGPT 4 app to record my own Science Pitch on a test basis and relatively spontaneously. This laid the foundation for a personal and seemingly authentic text. The AI then took out filler words such as "um", "yes" and "quite". Based on my instructions, it split particularly long sentences into several short ones. By doing so, you can take away two crucial advantages: spoken texts sound stylistically more active than written texts, and the text you have pre-written is based on your "self-produced and self-structured thoughts, which, however, were additionally enriched with ideas and information by the AI during the creation process" (Lang and Hütter 2023c, p. 38).

The prompt for a short presentation could read something like this: "Please create a Science Pitch of just under 3 min based on the following text. Rephrase the text so to eliminate unnecessary wording, filler words, and repetitions. I aim to inspire personally, to convince in terms of content, and to encourage my audience to engage in conversations through a personal approach. My target audience is professors, institute directors, and scientists. [Now your recorded text follows]."

ChatGPT edits the spoken text and adds words such as "fascinating", "reveal", and "amazing". You can use words that suit your style and that you usually say, but you should replace others with words that suit you and the context. Your spoken text should not sound stilted but natural.

You can develop your Science Pitch step by step using the questions in the book. Play your answers back to the AI as needed and look for suggestions for active wording. A prompt for this could be:

> Rephrase the following text into two short sentences—in such a way that it becomes an interesting short presentation of the project as part of a Science Pitch: "Founded in 2021, CarbonoLocal is a start-up aimed at small farmers in Latin America who farm in an environmentally friendly way. The CarbonoLocal team helps them join a grouped international carbon certification scheme and sell the CO_2 certificates in Europe to reward them for their small but significant contribution to climate protection. In their pilot project in San Juan de Rio Seco in Colombia, the founders of CarbonoLocal Nataly and Laura Cubillos also promote the development of a local bioeconomy and the decentralized production of renewable energy." (Weltweit 2021).

ChatGPT replies: "CarbonoLocal, established in 2021, empowers Latin American eco-farmers by facilitating their entry into a global carbon credit market. The pilot in Colombia not only trades CO_2 certificates but also pioneers local bioeconomy and renewable energy."

Sometimes, ChatGPT replies in rather long sentences. In this case, you can write a prompt like this to ChatGPT: "Thank you. That is still too long for me. Please write two short sentences with a maximum of ten words each."

This text can also be summarized in one sentence. My prompt to ChatGPT: "Rephrase the text from CarbonoLocal into a one sentence pitch. This sentence should contain a maximum of ten words and arouse curiosity."

ChatGPT replies: "CarbonoLocal: Transforming Latin American agriculture into a sustainable, global eco-force."

You can also ask ChatGPT for 10–20 formulation suggestions and develop your preferred wording from the most suitable proposals.

This approach will generate new ideas for phrases you can skillfully embed in your Science Pitch. Avoid pure "copy-paste", though: especially science is original and committed to the truth!

With an attractive-sounding presentation title (the one sentence pitch), you will attract attention and interest. With your Science Pitch, you demonstrate that you can skillfully summarize scientific facts. You can only hint at details for time reasons, so you will often remain vague. As a result, you will be caught between the demands of scientific precision and the need to limit the content you share. For this reason, AI can only supplement your work for now (Fig. 8.1).

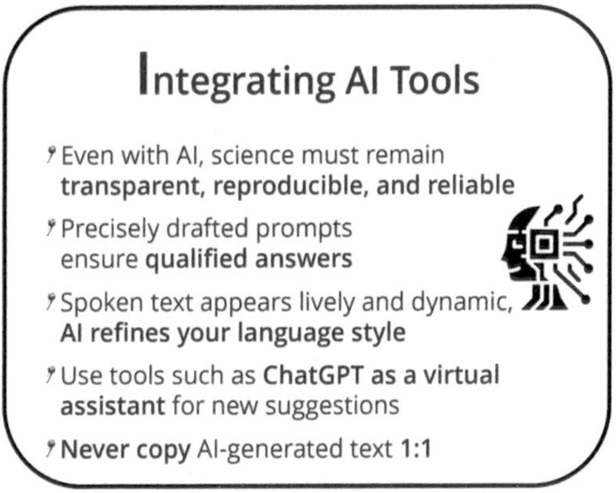

Integrating AI Tools

⸕ Even with AI, science must remain **transparent, reproducible, and reliable**

⸕ Precisely drafted prompts ensure **qualified answers**

⸕ Spoken text appears lively and dynamic, **AI refines your language style**

⸕ Use tools such as **ChatGPT as a virtual assistant** for new suggestions

⸕ **Never copy** AI-generated text 1:1

Fig. 8.1 Key Takeaway: AI tools such as ChatGPT are a valuable addition to preparing your Science Pitch (source: author illustration)

References

Erduran, Sibel 2023. AI is transforming how science is done. Science education must reflect this change. Expert Voices. Science, 382/6677, https://doi.org/10.1126/science.adm9788. Accessed 19 Jan 2024.

Frueh, Sara 2023. How AI Is Shaping Scientific Discovery. Feature Story. National Academies, Sciences Engineering. https://www.nationalacademies.org/news/2023/11/how-ai-is-shaping-scientific-discovery. Accessed 19 Jan 2024.

Lang, Sandra Mareike and Hütter, Franz 2023a. Serie KI im Training. Transformer der Weiterbildung. Training aktuell, 10.2023, 36-41.

Lang, Sandra Mareike and Hütter, Franz 2023b. Serie KI im Training. Die KI ist auch nur ein Mensch. Training aktuell, 11.2023, 38-43.

Lang, Sandra Mareike and Hütter, Franz 2023c. Serie KI im Training. Musenkuss 3.0. Training aktuell, 12.2023, 34-39.

Lang, Sandra Mareike and Hütter, Franz 2024. Serie KI im Training. Zwischen Chance und Risiko. Training aktuell, 12.2023, 36-41.

McKinsey 2023. The economic potential of generative AI: The next productivity frontier. https://www.mckinsey.com/capabilities/mckinsey-digital/our-insights/the-economic-potential-of-generative-AI-the-next-productivity-frontier#introduction. Accessed 19 Jan 2024.

Weltweit 2021. Nataly and Laura Cubillos: CarbonoLocal. Innovation and Certification for Local Climate Initiatives. https://welt-weit.org/en/project/carbonolocal/. Accessed 27 Jan 2024.

Present Your Slides with Power and to the Point

9

Abstract

Depending on the time frame of your Science Pitch, use slides very sparingly and do so in a targeted and particularly effective manner. This means that outlines, thank-you slides and lengthy literature references are generally left out. They do not provide additional value to your audience. Shift your references to a handout instead.

Many slides appear as data trash with overloaded text in extensive bullet-point lists and labels of complex figures. It disconnects the speaker from the audience. Instead, you can present your topic briefly and clearly by displaying only one essential piece of information per slide. Choose wisely what you highlight thematically. The guiding principle is "Less is more!"

You can reveal complex information step by step. A short interaction with your audience is often an excellent intermediate step before unveiling your results. Also, remember that graphics often convey more information than mere words!

"Visual illustrations play a crucial role in communicating scientific information." They contribute to better understanding and to forming public opinion. The decisive factor here is that you, as an expert on your subject, make sure it is easy to understand, even for your non-specialist audience. Complex information is, therefore, best presented sparingly and with little text (Gantenberg 2023).

Depending on the timeframe of your Science Pitch, use slides only very sparingly, in a targeted and particularly effective manner. You can abandon outlines, thank-you slides, and lengthy references. They do not add value to your audience anyway. You can include a list of references in the handout if required.

9.1   Less Is More: Just One Piece of Information per Slide

It is a common phenomenon that scientists and other experts want to share as much information as possible with their audience. This is understandable because they want to emphasize their expertise. However, if they change the perspective from expert to interested audience, they quickly realize that, for many people, this is overwhelming. The content of the slide must first be identified and understood. The audience often sees a dataset with extensive labeling of figures while the speaker is already talking about details. The audience is thus quickly left behind.

The following slide (Fig. 9.1) features several mistakes at once:

- *It is heavily overloaded with eight images.* A single image that covers the entire page would be preferable. In addition, microscope images should be displayed as enlarged as possible anyway.
- *The images do not reflect the natural sequence* shown on the right model. The spatial reference can only be understood by looking closely at the vertical sequence e11—e10—e8—e2.
- *It is a text-heavy slide.* Still suitable for a handout, the audience quickly gets lost during the presentation. The most important features are mentioned in the presentation and can be animated by arrows or markers in the photo. In addition, a small text font is readable mainly in the front rows only. If text is displayed, it should be easily readable from the last row.

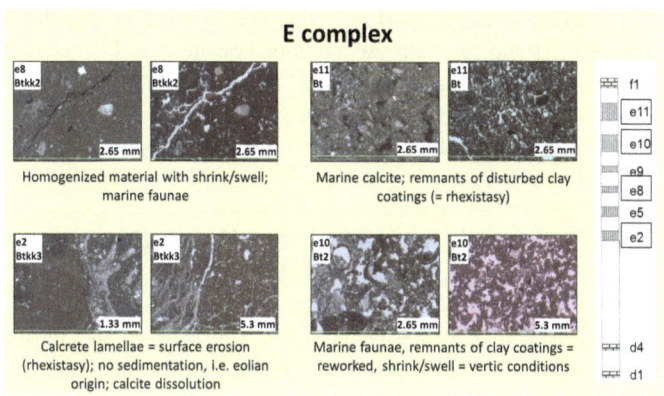

Fig. 9.1 Example of a poorly prepared PowerPoint slide overloaded with content (source: author illustration)

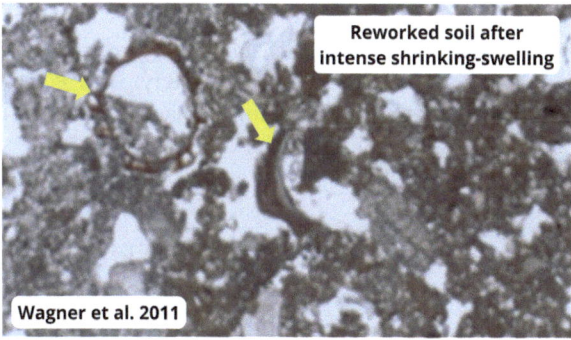

Fig. 9.2 Example of a high-quality PowerPoint slide that focuses on one key aspect only (source: author illustration)

Limiting yourself to the two or three most important images is much better. In a time-limited Science Pitch, even that can be too much. Always ask yourself: what does the image show? Can your audience grasp the content you convey? Does your slide add to your take home message?

Figure 9.2 shows a better alternative to Fig. 9.1:

Here, you could initially show the image without text. If presentation time allows, ask your audience what they recognize. This creates an additional opportunity for interaction. You respond to their answers and can then use a unique key message ["Reworked soil after intense shrinking-swelling" in Fig. 9.2.] to resolve what you want to pass on. This way, your presentation becomes understandable and comprehensible, and you get to the point.

9.2   Essentials Instead of Text-Heavy Data Trash: Get Rid of Bullet Points!

Many speakers display text-heavy slides. This is based on a misunderstanding: unlike Word, programs such as PowerPoint, Prezi, and Keynote are not intended for documentation but for presentations.

If I hand you a fork and a knife and ask you to eat the tomato soup I have just served you, you will still sit at the table when the soup is already cold. Do not blame the knife because you are still hungry. You do not have to cut the knife through your soup!

The same applies to PowerPoint and Word. If you use both correctly, you can communicate very effectively: with PowerPoint for your presentation, with Word for the handout, and for documentation and publications (Zimmer 2023).

Figure 9.3 shows a slide that is certainly not suitable for a speech:

The Most Common Mistakes in Visual Presentations

- Full cover slide with title, name, organization, sponsors, photo, contact information, table of contents, and other details.
- Include details: Lots of text, location descriptions, and references.
- Show columns of numbers in a large table (only three are important).
- Show 12 photos with detailed text descriptions – all on the same slide!
- Mention the conference title, venue, and sponsors on each slide (so your audience knows where they are).
- Your name and the title of your presentation belong on every single slide. Be sure to include details such as logos, sponsors, etc.!
- Include as many bullet points as possible on all slides.
- Read out your text in a different wording. Use the pointer to show where you are. Your audience needs orientation!
- Remember the "Thank you for your attention!" slide. Important: Mention 27 friends and colleagues individually and show a group photo.
- Show a photo of your favorite dog lying comfortably in the garden.

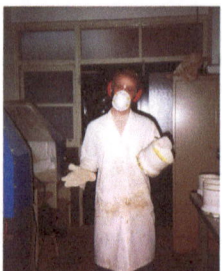

Dos and don'ts with visuals in presentations – Conclusion: The most common mistakes in presentations
Always state your name and as much additional information as possible. Share as much content as possible with your audience!

Fig. 9.3 Example of a PowerPoint slide overloaded with text (source: author illustration)

You see ten key points you can include in your handout—just not on your slide. Otherwise, your audience must decide whether to listen to you or read along. Some will switch off immediately and move on to other things, such as the latest news on their favorite social media channel.

Instead, if you know the most important aspects, you can reproduce them immediately. It is not a matter of reproducing every aspect perfectly, but only the essence. Your presentation time is limited, especially in the Science Pitch. The viewer will also not understand the connection between the photo and the topic of the presentation.

In practically every workshop, I see scientists and experts displaying the logo of their company or research institute on every single slide for branding reasons. They usually explain that this is mandatory and standard. From a mere presentation perspective, it wastes valuable space so that the photos, graphics, and sketches are reduced in size by up to half. This takes away much of their impact, making some essential details challenging to recognize. If you instead gain a sense of what is essential, you can selectively choose which of the depicted aspects you want to highlight or discuss.

Compared to Fig. 9.3, Fig. 9.4 looks much more organized: here, your audience can see at a glance what the following aspects

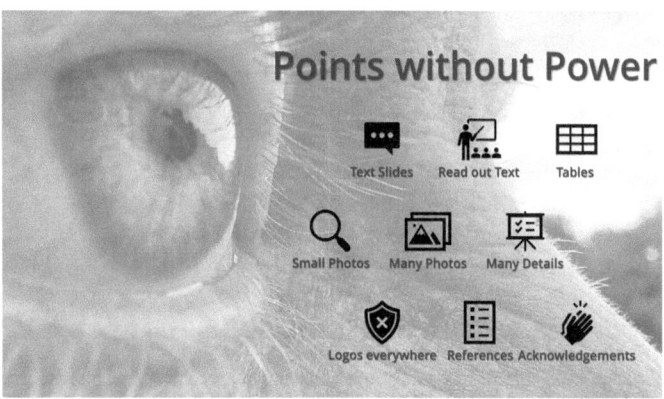

Fig. 9.4 Example of a neat, well-organized PowerPoint slide (source: author illustration)

are about. Now, just select the one or two most important ones and embed them in specific practical examples in your speech.

9.3   Visualize the Story Behind Your Statistics

What is the best way to visualize statistical data? Certainly not through lengthy columns of numbers in written and tabular form that your audience must work their way through. Present only the data that matters! Figure 9.5 shows an intermediate step in which the length of the bar charts represents percentages.

It is more attractive and entertaining if you visually emphasize statistical data even more, as in Fig. 9.6:

In both cases, you can reveal the information step by step: in Fig. 9.5, you start by displaying only the bars and ask your audience for possible answers to your initial question. Then, it is about time to reveal the actual results. You can also compare these with other survey results. Figure 9.6 visualizes the four answers, while the grey or colored background also indicates the percentages. You can display the numbers after a corresponding interaction.

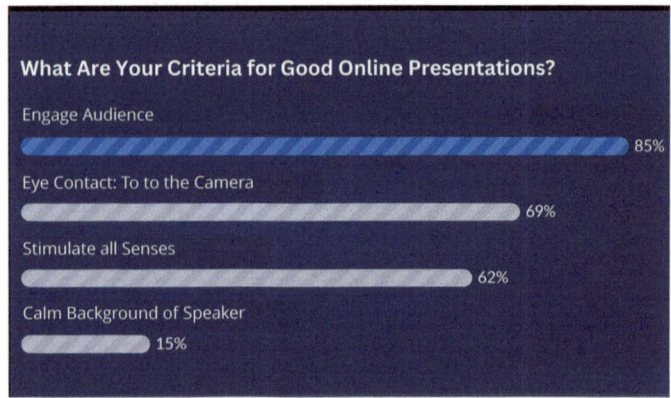

Fig. 9.5 Example of a conventional presentation of survey results (source: author illustration, based on an online survey with slido.com)

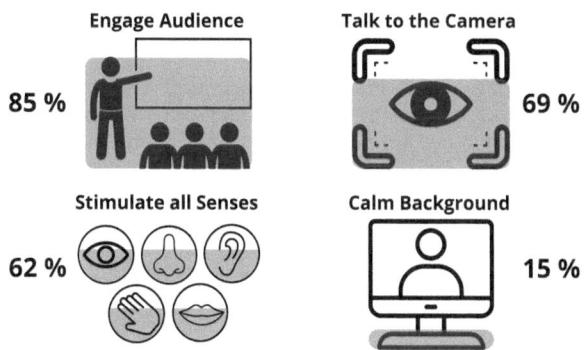

Fig. 9.6 Example of a visually appealing presentation of survey results (source: author illustration)

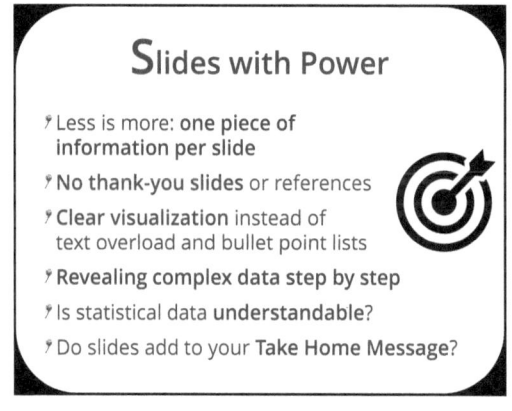

Fig. 9.7 Key Takeaway: How to present with power to the point (source: author illustration)

The results presented here are based on a short survey in an online audience of 13 people in April 2020. They do not claim to be representative and only reflect trends, which can vary greatly depending on the individual case. The example shows that such prepared information is more appealing and interactive than simple table data.

In a nutshell, Fig. 9.7 highlights major aspects of a successful presentation with slides.

References

Gantenberg, Julia 2023. Eine Grafik sagt mehr als tausend Studien. Gastbeitrag Wissenschaftskommunikation.de, https://www.wissenschaftskommunikation.de/eine-grafik-sagt-mehr-als-tausend-studien-72991/. Accessed 13 Jan 2024.

Zimmer, John 2023. PowerPoint ≠ Word!!! https://www.linkedin.com/feed/update/urn:li:activity:7143547558114594816?utm_source=share&utm_medium=member_desktop. Accessed 13 Jan 2024.

Prepare Your Setting Professionally

10

Abstract

You can professionally prepare a high-quality presentation. The better you are prepared, the more confidently you can handle spontaneous changes. Familiarize yourself with your presentation venue early on and get to know the setting for optimal stage presence. Conduct a technical check and ensure that you can present without technical aids in case they are not available or fail at short notice.

A high-quality technical setup is also essential for online presentations. You should be well-prepared, from a microphone that ensures a clear and understandable voice to good lighting and a camera that makes you visible. Arrange your background so that it does not distract from the presentation. Consider wearing clothes in colors that match your background. Stand up during your presentation for a more dynamic delivery rather than sitting down. Lastly, position your camera at eye level and look directly into the camera lens to establish virtual eye contact with your online audience.

Record your Science Pitch as a test to analyze and gradually improve it using voice and video analysis. You can get yourself into the right mood on the day of the presentation and just before it.

Your presentation will only convince your audience if you appear professional. This includes equally professional preparation. Whether your audience sits next to you in person or if you are presenting online or in a hybrid format, ensure that you convey a consistent image. Even if you cannot plan every detail and must react immediately to spontaneous changes, targeted preparation will help you.

10.1   Optimum Stage Presence for Your Live Talk

Do you know the room where you will deliver your Science Pitch? Familiarize yourself with the stage setting early on. Whenever I attend conferences or events where I am going to give a speech, the very first action I take is to inspect the presentation room.

Pay special attention to these questions:
- How is the stage set up?
- How much space do I have to move around?
- How many people can I expect at the event?
- How is the seating arrangement?
- Where can I find the best audience connection?
- How close will I come to the audience?

In traditional lecture halls, the distance is unnecessarily increased by large tables, the lectern, the (mobile) video projector, or even the overhead projector. Keep in mind that closer physical proximity can facilitate audience connection, thus allowing for more direct and personal communication. Can you even walk through the rows or table groups?

Physical presence and clear visibility to your audience are crucial for personal and convincing communication. I once experienced a parent-teacher conference at my daughter's high school, where a teacher stood almost in the corner of the assembly hall. For those facing her directly, this might have been somewhat acceptable; however, I was sitting on the other side and could only see her very small and barely hear her. You should take as central a position as possible, especially at the beginning and end of your presentation.

The use of headsets and microphones is standard in larger rooms. Get in touch with the technical staff well in advance. I have had good experiences with acoustic tests. This involves questions like "What position of the headset or microphone produces the optimal resonance in the room?" and "Is the headset comfortable and stable, without shifting?". The headset should not cover your face. To reproduce your voice, it should be close to your mouth but not overly emphasize every sound, like breathing in and out. If you have long hair or wear jewelry (e.g., long earrings), ensure they do not create unnecessary background noise. Also, the cable should not pull uncomfortably.

Check the optimal position of the microphone in advance. This way, you avoid the mistake many speakers make without preparation: holding the microphone too far away from the mouth, making their voice unclear, or too close, causing crackling sounds. Be more cautious with expansive gestures, and keep the microphone between your chin and mouth while speaking. If you get too close to the speaker, there can be unpleasant interferences with extraordinarily high and uncomfortable noise.

In addition, pay attention to optimal lighting conditions. You should be well-lit and thus always visible to your audience, but also be mindful of the natural daylight coming from the side and the changing conditions depending on the weather.

If you bring your laptop, do you need power, an extension cable, and a connection between the laptop and the projector? What equipment is provided on-site? Bring your HDMI connector. In the control panel, you can set your laptop so it does not go into sleep mode after a few minutes. Use "Control Panel—Hardware and Sound—Power Options—Edit Settings". Especially if you are not a tech enthusiast, arriving at the event early is always a good idea. This allows you sufficient time to check and adjust all technical details, avoiding unnecessary interruptions during the presentation. Ensure that your laptop's battery is fully charged.

Most presentations are supported with software or apps like PowerPoint, Keynote, or Prezi. Ensure that you can open and use your file. I always bring my laptop and a USB stick, have the file saved in my cloud, and often send it via email to the organizer and myself. But what if the projector's bulb fails and a replacement cannot be found quickly? Then, you should be able to present your Science Pitch spontaneously with the support of a whiteboard or flipchart—or even without any aids. Test these variations thoroughly in advance. Switching your media confidently in unusual situations demonstrates your high level of professionalism, and your audience is more likely to forgive minor mistakes.

You can show your images and graphics for small groups even directly from your laptop or tablet. You might have printed the slides and can refer to them now. It is best always to prepare so you can present without aids if necessary.

10.2   Your Virtual Business Card: An Awesome Online Presence

When presenting your Science Pitch to an online audience, some critical prerequisites for a successful presentation change. Technical preparation becomes even more crucial, especially for voice and video transmission, that is, audio and video.

The most important is a clear, distinct voice. You should speak somewhat slower than in front of a live audience, as high speeds occasionally lead to sound distortions. A varied voice is particularly important online because it keeps your audience engaged and fosters their attention.

Make sure you use a high-quality microphone. Internal laptop microphones often have an inferior sound quality compared to external microphones. Disturbing background noises, especially keyboard noises, are transmitted quite clearly. You can significantly improve the sound quality with a headset, lavalier microphone, or USB microphone. A headset reduces ambient noise and ensures good sound quality as the microphone is close to the mouth. On the other hand, not everybody is comfortable; it can also detract from your appearance. A lavalier microphone is much more subtle and has a sound quality comparable to a headset. Ensure the cable fits tightly and does not shift with extended movements. In my opinion, USB microphones have the best sound quality: the integrated pop protection makes sizzling noises disappear; you can also adjust the sound optimally. On the other hand, you need to be close to the microphone for a consistently stable sound, and USB microphones cost more than other types of microphones. In conclusion, a stable and clear sound is crucial. Your audience should always be able to understand you with ease.

A pleasant screen background is essential to ensure high-quality video is transmitted. What is visible behind you? A neutral, solid-colored background is sure to be the least distracting. You can add depth to your virtual room by aligning the vanishing lines to converge towards the back. Avoid unnecessary additional movements. For example, in one of my online trainings, a participant had placed a Maneki-Neko, a Japanese "waving cat", in an otherwise very tidy background. It was challenging to concentrate on the content of her presentation for the next 10 min. When online meetings started popping up in early 2020, I often looked at cluttered bookshelves. Curious people then tend to look more closely at which books are on the shelf. Extreme cases included ironing boards and laundry racks, untidy bedrooms, garages (!) with clutter, garden sheds, and photos and posters of various sizes. A calm background works best because the focus remains on the person and the presentation's content

Make sure that your face is well-lit. Daylight should not come from behind but ideally from the front or at least from the side. You can use additional light from the front, side, or diagonally from above. Additional softboxes take out the harsh light.

What clothing will you wear? Ideally, it should be in distinct color contrast to your natural or virtual background. But be cautious: checkered, striped, or other patterned shirts and sweaters flicker on camera and, therefore, appear particularly restless. This is why I prefer solid-colored clothing. An extreme case I experienced was an online participant dressed in a shirt on top but only wore underwear below. This was quite awkward when he stood up and walked around. Even though we usually only see the upper body online, I always wear comfortable shoes, so I feel at home as if standing on a real stage.

By the way, I recommend delivering your presentation standing rather than sitting. This adds a higher level of energy. When sitting, your body slumps down more quickly, meaning you must straighten up repeatedly. Do not look into the camera from above, below, or even from the side, but always at eye level. It makes a huge difference to look directly into the camera lens, not at your screen (Fig. 10.1). Even while sharing slides with PowerPoint, you should still convey your most important messages through direct virtual eye contact.

10.3   Get in the Right Mood for Your Presentation

You can test your presentation beforehand by recording and analyzing it: first, listen to your voice and the content you are presenting. This will raise awareness of the highlight moments and the aspects you want to improve. Then, do the same in the second round by analyzing the video (without sound). Next, combine sound and video and decide on the two or three most important aspects you want to work on. Repeat this until you are satisfied with your performance.

Prepare your Science Pitch not only in terms of content. On the day of the presentation or immediately before, you can get into the right mood by mentally walking through the setting and scenario: where will you be standing? Who will be listening to you? What expectations can you anticipate? It can also help if you have a good night's sleep and eat something good and not too much. Relaxation exercises or your favorite music might also be beneficial. I often imagine my audience hanging on my every word when I present certain content. I also recall particularly positive experiences from the past.

> # Stage Setup
>
> ⸙ Familiarize yourself with the **stage setup** early on: Layout, space, audience contact
>
> ⸙ **Technical check**: Microphone or headset? Light? Screen transfer etc.
>
> ⸙ **Online**: Sound quality, present while standing, camera at eye level, look into the camera lens
>
> ⸙ Ensure **mental hygiene**, positive mood, optimal preparation

Fig. 10.1 Key Takeaway: Prepare an appealing stage setting for your Science Pitch (source: author illustration)

Part II

Science Pitch: How to Add ESPRIT to Your Project

Now let's move on to the core of your Science Pitch, the ESPRIT. We will discuss your professional expertise and how you build your storyline. Based on your intrinsic values, we focus on the current research question in your project and your approach to solving it. With your performance, you can present yourself authentically and showcase your passion for your project. Then we delve into the content details: you highlight your project's relevance and innovative nature. You also address its feasibility and potential knowledge transfer. At the end of your short presentation, you condense everything said into one or two brief sentences. This provides your audience with a memorable message: your take home message.

The Meaning of ESPRIT

11

Abstract

A successful Science Pitch is delivered with ESPRIT. It combines wit, brilliance, and intelligence, focusing on the essentials of your research. Within 3 min you express your full personality elegantly combined with your expertise.

When all factors work together consistently and suitably, you present your Science Pitch with lots of **ESPRIT**, as highlighted in the video (Fig. 11.1). According to Duden (2024, translated), Esprit means a "witty, brilliant [way of speaking] that sparkles with intellect and spirit". Witty because you have thoughtfully considered in advance which aspects of your project bring the most value and because you can focus on the essential content. Brilliant, because your audience immediately understands the primary focus and experiences a special aha moment. Sparkling with intellect and spirit by showing technical brilliance, presenting charmingly, and sharing your passion for the topic. Creativity and

Supplementary Information The online version contains supplementary material available at https://doi.org/10.1007/978-3-658-44844-8_11. The videos can be accessed individually by clicking the DOI link in the accompanying figure caption or by scanning this link with the SN More Media App.

Fig. 11.1 Prepare your talk by applying the ESPRIT model to your Science Pitch Canvas (source: author illustration) (▶ https://doi.org/10.1007/000-cz0)

a touch of humor round off your Science Pitch. It is crucial to present the essentials in a maximum of 3 min. Express your entire personality and combine it elegantly with your expertise.

Reference

Duden 2024. Esprit, der. Cornelsen Verlag GmbH. https://www.duden.de/rechtschreibung/Esprit. Accessed 25 Jan 2024.

Expertise: Combine Innovation and Personality

12

Abstract

A convincing Science Pitch is based on your verifiable expertise. Therefore, share relevant practical experience that provides recognizable added value to your target audience. You support your expertise with robust and fact-based information. This contributes to your competence and credibility and stands for your objective stance. However, be aware that technical excellence without passion for the subject is of limited value; if you can combine both, that is half the battle for your success.

Build a bridge for your audience by inviting them to a dialog following your presentation. You can also demonstrate your expertise by translating abstract information into relatable, immediately understandable terms. Emphasize what changes through your involvement in the project and what crucial impetus you provide for the further development of existing knowledge.

The better you can measure and scale your innovation, the more convincing your Science Pitch will be. Factors such as its long-term impact on research and practice are valuable hints. It is also about your personality, i.e. how well you identify with your topic and what skills you bring to the table. In a narrower sense, these are your verifiable references and patents and how diverse and valid your professional expertise and qualifications are.

12.1   Highlight Your Professional Expertise

Does your idea match your audience's interests, prospective cooperation partners, and potential customers? What advantage do
they have if they support your idea instead of others? Also, answer
which of your specific skills set you apart from colleagues inside
and outside your field of research. In your personal unique selling
proposition (USP), you combine your character traits with your
professional qualifications.

Get more involved here than usual. By merging your project
and personal experiences, you ensure a distinctive presentation
and establish a personal connection with your audience. Do you
share common values, life experiences, or a particular passion
with your audience?

Peter Karacsonyi successfully implemented shared
experiences and passion in the TV show "Die Höhle der
Löwen" (the German version of the British TV show "Dragons' Den") in 2015: with his product "KAPE Skateboards",
he brought together particularly durable skateboards made
of carbon and fiberglass with high-quality natural woods
such as maple and bamboo. The high-quality skateboards
also stand out thanks to their particularly cool design. Three
of the five jurors, or "lions", were out immediately despite
the excellent performance because they did not relate to
skateboards. The two other judges made concrete offers,
with Karacsonyi awarding the contract to investor Frank
Thelen. Why? Because Thelen used to be an enthusiastic
skateboarder himself and could immediately identify with
the advertised product! (VOX 2015)

So first, consider who you are dealing with, whether your target group feels connected to your topic or idea, and what could
inspire them. This is how passion and excellence are brought
together. For Dr. Larry Smith, Professor of Economics at the

University of Waterloo in Canada, passion is key to innovation: "If you are to have a great career, you must be an innovator and you cannot innovate without passion" (Gallo 2016). Without personality, you can also be represented by a robot or artificial intelligence (Wagner 2023).

A Science Pitch is much more than just a brief description of your project.

Engage your audience, share your enthusiasm, and underline your qualifications:

- What makes me a qualified researcher?
- Do I know my inner strengths?
- What am I passionate about in my project?
- My USP: what characteristics, qualifications, and experiences are unique to me?
- How do all these facets contribute to the success of the project?

The Science Pitch does not involve mentioning every qualification. Choose wisely which ones are ideally suited to the specific occasion and your target audience.

In my scientific career, I was the head of a laboratory at an institute for four years. However, this experience was irrelevant when I took on a position as a career coach. My long-standing presentation experience and empathy in coaching were the crucial skills I brought to the interview ultimately leading to the job offer.

When you anecdotally describe personal experiences, you make yourself unique and, therefore, distinctive. This also prevents intellectual theft because nobody can copy your personality

1:1. At the same time, you showcase your specific strengths and create a stable basis for your success.

In addition, your sponsor first invests in you as a personality (Fig. 12.1). Only then comes your project idea. Keep this in mind if you are applying for funding via grants or if you start a spin-off from the university.

Invite your audience to engage in a dialog. This can be in a discus-

---

Once you have thought through all the steps, add an emotional key message:

- How do I want to activate my audience at the end?
- Do I ask them to take a specific action?
- Do I convey a memorable take home message that will change their attitude?
- How do I want to be remembered? What do I want people to remember?

---

sion round immediately afterward or as part of the event. It is a valuable aspect of your presentation after the Science Pitch, as you can demonstrate your expertise with your answers to audience questions.

12.2 You Convince with Sound Data and Reliable Facts

Emphasize the general social benefit of your project. You should briefly describe the extent to which your research can have an economic, ecological, cultural, social, or political impact. You can refer to the impact of previous projects and publications for a better assessment.

You achieve a personal impact primarily by emotionally connecting with your audience. Can you substantiate your theses and previous results? It is also about vividly communicated data, facts, and figures. Solid facts reflect your objective stance and satisfy your critics making you appear both competent and credible.

You prepare your data, facts, and figures as visual, memorable, and structured as possible: transform tables into diagrams, focus on the essentials, support your statements with photos and metaphors, and prioritize your content. The result will be a dynamic presentation (Restle 2022).

You need sound information with a solid database to convince your audience of your content. Regardless of the precision of your data, it is crucial that your audience immediately understands it.

Emma Horn, PhD student at the University of Cape Town in South Africa, gives an excellent example at the Falling Walls Lab 2022 in Berlin: "Ceramic tiles: 16 billion square meters were produced globally last year. That's enough to tile the entire area of Berlin almost twenty times over." In this way, an abstract parameter becomes a concrete idea, even regarding the location of the talk.

In the same way, your methodology should be reasonable.

Deepika Kurup (2016) provides a vivid example in her TEDWomen talk. She presents photocatalysis as a newly used method for water purification: "So what exactly is photocatalysis? […] "Photo" means from the sun, and a catalyst is something that speeds up a reaction. So, what photocatalysis is, it's just speeding up this solar disinfection process [...] to remove bacteria and organics and a whole lot of contaminants from drinking water." Kurup underpins the information and the methodology for removing the pollutants with visual animation. In this way, the scientific content of the presentation is conveyed and understandable, even for those who are unfamiliar with the subject.

Your hands-on and easy-to-understand embedded facts and figures are essential evidence of your expertise and the impact of your project: what will change? Which details and particularly meaningful results will provide the decisive impetus for change and progress? The greater the potential for practical application and scaling of your innovation, the more valuable your expertise will be.

There are further opportunities in the knowledge transfer and social benefits. This extends far beyond pure data, facts, and figures. You should also talk about the potential impact, i.e. the potential medium and long-term influence of your project on further research. At the same time, think about your impact on your presentation.

To accentuate your presentation, answer the following questions:

- What project results and personal stories will add spice to my Science Pitch? What does my audience identify with?
- Do I present my information thoroughly yet quickly, understandable and comprehensively?

12.3   Publications and Patents Are Your Solid Assets

Highlight your expertise by answering the following questions:

- What papers did I already publish?
- How often am I cited? What is with my h-index?
- Did I already publish in "Nature", "Science" or other high-ranking scientific journals?
- Did I already develop a patent?

Expertise: Personality

⚜ **Individual characteristics** combined with your professional qualifications.

⚜ **Passion before excellence.**

⚜ **Your sponsor** invests in your personality first, in the project idea second.

⚜ **Solid facts:** Objective attitude, competence, and credibility.

Fig. 12.1 Key Takeaway: Emphasize your expertise and personality (source: author illustration)

References

Duden 2024. Esprit, der. Cornelsen Verlag GmbH. https://www.duden.de/rechtschreibung/Esprit. Accessed 25 Jan 2024.

Gallo, Carmine 2016. The Secret to A Great Career and A Great Presentation. https://www.forbes.com/sites/carminegallo/2016/04/30/the-secret-to-a-great-career-and-a-great-presentation/. Accessed 09 Feb 2024.

Horn, Emma 2022. Green Ceramics: Breaking the Wall of Tile Manufacturing. Falling Walls. https://www.youtube.com/watch?v=P7Z-0HfKIzI&t=4s. Accessed 16 Nov 2024.

Kurup, Deepika 2016. A young scientist's quest for clean water. TEDWomen 2016. https://www.ted.com/talks/deepika_kurup_a_young_scientist_s_quest_for_clean_water. Accessed 14 Jan 2024.

Restle, Viola 2022. 30 Minuten Zahlen lebendig präsentieren. GABAL-Verlag GmbH Offenbach, 96 pages.

VOX 2015. Die Höhle der Löwen: Frank Thelen hebt mit dem Skateboard ab. https://www.vox.de/cms/die-hoehle-der-loewen-frank-thelen-hebt-mit-dem-skateboard-ab-2419135.html. Accessed 14 Jan 2024.

Wagner, Stephen 2023. Do You Present with Personality and Passion? https://www.youtube.com/watch?v=fAy2CUfOmkg&t=43s. Accessed 14 Jan 2024.

Storyline: How to Guide Your Audience with a Science Pitch

Abstract

With your vision, or the vision of your project, you set a future perspective. You talk about your motivation and specific intention, explaining what drives you in your project. The mission—your guiding principle—describes the path to implementing your vision. You then articulate the existing problem and a concrete solution approach, along with the key milestones of your research project. You should also gain clarity about your value system and the values you or your project represent.

Develop an arc of suspense with surprising twists. Quickly implementable story structures like the "Narrative Spectrum" and the "Pixar Pitch" provide valuable inspiration through practical examples. Ensure a crystal-clear message that you convey to your audience. To achieve this, translate your project's numbers, facts, and figures of your project into compelling stories with close practical relevance so your audience can always follow along. Like publications, Science Pitches are also truthful information. However, they are always compact, concise, and often abbreviated because they get to the point.

There is a simple and immediately understandable structure you can apply to present your Science Pitch successfully. You will convince and inspire your audience if you share passion and expertise with authentic speech delivery. According to Frank

© The Author(s), under exclusive license to Springer Fachmedien Wiesbaden GmbH, part of Springer Nature 2024
S. Wagner, *Science Pitch*,
https://doi.org/10.1007/978-3-658-44844-8_13

Asmus, citing Friedemann Schulz von Thun (2021, p. 21), "Authenticity is more important than rhetorical effectiveness".

13.1   Vision: A Purposeful Look into the Future

Do you know your vision or at least the vision of your project? Your answer to this question is important because it looks into a future that does not yet exist. It focuses on the specific intention of why and for what purpose you are doing something.

Your vision is emotionally charged and moves you into action. Visionaries imagine their distant future in very concrete terms. The Wright brothers' vision was to launch an engine-powered airplane. In 1903, the time had come. In 1961, US President John F. Kennedy set the goal of landing on the moon at the end of the 1960s. A not exactly trivial, but achievable goal that inspired many people. A vision is not easy to realize. In 1962, Kennedy said: "We choose to go to the Moon in this decade and do the other things, not because they are easy, but because they are hard …".

Your vision is formed in your head and corresponds to the factual level.

> Take your time to answer the following questions:
>
> - What higher goal am I pursuing in my life?
> - Why does my project exist? What is the purpose of this project?
> - Why am I involved? What do I want to achieve?
> - What could reality look like once the vision has been realized?

Purpose addresses your intrinsic motivation, i.e. why you are doing something and what it is for. It is also part of your identity. Simon Sinek explains this impressively in his well-known TEDx talk (Sinek 2009) and in his book "Start with Why" (2011).

> Your purpose corresponds to your goal, your concern, or your conviction and answers the following questions:
>
> - What belief systems and ideas are essential to me?
> - What motivates me to work on my current project every day? How does this fit in with my personality, experience, and expertise?
> - Why should the outside world care about my project?

According to Sinek (2009, 2011), if you recognize the more profound meaning in your work, you will also get through the difficult times in your project: you should not give up on your work after just a few failures but accept challenges and find new ways to solve problems and achieve goals.

If you can answer the above questions and summarize them in a few short sentences, you will have developed a valuable component for your Science Pitch and many personal discussions. Table 13.1 provides real-life examples of visions or vision statements.

Table 13.1 Examples of visions—or vision statements—for companies, associations, and projects

Name/organization	Vision/vision statement
Google (Google Bard 2024)	Google's vision is to make the world's information universally accessible and useful.
DFG (DFG 1955, 2023, p. 1)	The DFG promotes research of the highest quality.
DAAD (DAAD 2024)	Change by exchange.
Project ManuBrain (Fraunhofer SCAI 2024)	Universally applicable, scalable AI platform for industrial applications. (translated from German; AI: artificial intelligence)
Dr. Rainer Wild-Stiftung (2024)	Promoting research and development of healthy human nutrition. (translated from German)
My Vision	People and experts of all different backgrounds communicate at eye level and inspire each other.

13.2　Mission: Turn Your Vision into Reality

Your mission or mission statement describes the path to implementing your vision. It corresponds to the "How", i.e., how you implement your vision in everyday life and coincides with your feelings. Your heart is representative of this.

To frame your mission, answer the following questions:

- What drives me every day?
- What is important to me personally?
- What intrinsic motivation do I associate with my research and project?
- What deeper meaning do I see in carrying out my project?
- What emotions connect to my project? What excites me about it?
- What contribution, what tangible added value am I making with it?

It is all about relationships and what you achieve through your mission. The shorter and clearer your mission is, the more effective it will be. Larger organizations such as the German Research Foundation (DFG) comprise several levels and formulate more missions accordingly (Table 13.2). Formulate your mission positively instead of emphasizing what you do not want.

Table 13.2 Examples of Missions for companies, associations, and projects

Name/organization	Mission
Google (2024a)	Our mission is to organize the world's information and make it universally accessible and useful.
DFG (1955, 2023)	Focus on knowledge-driven research. Funds research projects, and conducts review, evaluation, selection, and decision of research proposals. Shapes standards of academic research. Maintains dialog with society, politics, and business and supports the transfer of knowledge. Advises state institutions and institutions working in the public interest on issues relating to academic research and research policy.
DAAD (2024)	The DAAD stimulates internationalization and global responsibility. It acts as an independent intermediary between various partners and their interests. The real motors of change, however, are students and researchers who cross borders every day through their work, and the institutions of higher education which support them in their commitment.
Project ManuBrain	Development of a universally applicable, scalable, and open AI platform specialized in industrial use cases to strengthen small and medium-sized enterprises in NRW.
Dr. Rainer Wild-Stiftung (2024)	Support health-promoting nutritional behavior in industrialized societies, develop solutions to counteract nutrition-dependent diseases, and strengthen health-promoting behavior. (Translated from German.)
My Mission	You know your strengths and present yourself confidently. You communicate complex information in a way that is easy to understand yet practical, entertaining, and always with a personal touch. You can easily present at the interface between science, business, and international affairs. You are recognized as an expert on your topic. I take your science out of the blackbox and into the lives of your audience!

13.3 Problem and Solution Approach: Milestones in Your Research Project

Briefly introduce your topic by describing the current problem. You do not have to disclose your project's sensitive or "secret" details here. It is usually sufficient to outline your solution approach briefly.

According to Simon Sinek (2011), this is about the "What", i.e., what you are doing and developing.

Answer these questions:

- What specific actions are associated with my project?
- What am I developing or improving through my project?
- What specific results should my project lead to?

Mention the most important milestones: what intermediate targets are you striving for in the individual project phases? Do you integrate these into your solution approach?

If you achieve your milestones, this will act as an additional motivation booster. It raises your awareness of the steps you achieve in each phase. Otherwise, you will hardly notice the progress.

13.4   Values: The Foundation of Your Daily Actions

What values do you act on every day? Which values would you never violate?

You can further distinguish between the values you stand for today and those you will strive for in the future. It is essential to clearly define each value for yourself or your project.

Table 13.3 provides an exemplary overview of values for companies (Martins 2023):

Table 13.3 Examples of values for companies and entrepreneurs

Name/organization	Values
Asana (Martins 2023)	To achieve great things—and quickly. Clarity. Co-creation. Assume and transfer responsibility. Mindfulness. No false compromises. Being authentic (with yourself and with others). Warmth.
Google (2024b)	Focus on the user and all else will follow. It's best to do one thing really, really well. Fast is better than slow. Democracy on the web works. You don't need to be at your desk to need an answer. You can make money without being evil. There's always more information out there. The need for information crosses all borders. You can be serious without a suit. Great just isn't good enough.
LinkedIn (LinkedIn Careers 2024)	We put members first. We trust and care about each other. We are open, honest, and constructive. We act as One LinkedIn. We embody diversity, inclusion, and belonging. We dream big, get things done, and know how to have fun.
Netflix (2024)	Judgment, Selflessness, Courage, Communication, Inclusion, Integrity, Passion, Innovation, Curiosity.
My Values	Honesty and Sincerity, Respect and Appreciation, Curiosity and Ambition.

To illustrate this by example, let me describe my definition of "curiosity": I am open to new experiences, interested in new developments, and trying out new things. In practice, this was the rapid switch from live to online presentation training at the beginning of 2020. In 2021, I supplemented my daily practical experience with professional training to become a certified online trainer. My customers appreciate the consistently high level of presentation training. In 2023, my curiosity led me to visit a new cooperation partner in Cameroon within a few weeks, delivering speeches, training, and coaching to hundreds of people in Yaounde, Douala, Bamenda, and Buea. A trip that had not been on my agenda shortly before but which resulted in incredibly impressive and unique new experiences. My curiosity always leads me to unusual and valuable experiences that broaden my horizons and enrich my life.

Do not limit your values overview to a simple list; define and describe each value clearly.

You can take it one step further and answer the following question:

- Which values do I wish to integrate even more into my everyday life in the future?

Future values I would like to develop further include courage, vision, focus, internationality, people, and joy.

For my future, the value of "internationality" means that my feelings and thoughts are essentially and sustainably based on my previous trips abroad on six continents. In the future, I will pursue this to a greater extent by traveling more, allowing me to gain new experiences. I would like to exchange ideas with people worldwide more than ever before. I still have regular contact with people worldwide if I am not traveling. So, I like to attend international events again and again. I also spend much time on LinkedIn for networking and informal exchanges.

In the Science Pitch, I share specific experiences or actions from one of my previous stays abroad as an anecdote. This anecdote should fit well within the content and not break up the presentation. This is how I met my trainer and speaker colleague from Cameroon in the Virtual Speakers Association International (VSAI) network. During my 11 days in Cameroon, I performed eight times on different stages and was interviewed live three times by various TV stations (Wagner 2023).

In addition to your most important values, you should consider which values you want to distance yourself from. They represent a clear taboo for you.

I distance myself from these values: hate, intolerance, hasty prejudices, personal insults, and ignorance.

I do not coach anyone who communicates condescendingly or repeatedly insults other people. This is where I benefit from my instinct, which rarely lets me down. I can sense very quickly whether the person I am talking to shares my most important values or represents my no-go values. Instead, I always like to coach at eye level.

"Hasty prejudices" correspond to classic stereotyped thinking: I know people who categorize others into certain groups before they even know each other personally. This deprives them of the unique opportunity to remain open-minded and curious and to learn from each other. Other people may end up in categories they may rarely feel they belong to.

13.5   Create Aha Moments for Impactful Messages

There are numerous storytelling structures. The best known are the classic hero's journey with the monomyth by Joseph Campbell (1949) with its 17 elements and the hero's journey according to Christopher Vogler (1992) with 12 elements. However, these two schemes are not suitable for a Science Pitch. Instead, I will introduce two easily applicable storytelling structures for your Science Pitch that reach beyond the usual introduction—central part—conclusion model. You can use an arc of suspense with breaks and twists for your story, in line with the previously mentioned sequence of situation—problem—solution—insight—outlook. This will ensure your audience listens to your presentation without getting lost in thought.

You can follow the path of your hero's journey based on the description by Browning (2021): you are a scientist and have a thesis that you have spent three years trying to prove or disprove. After overcoming several challenges, you have your aha moment: you have found the answer to your research question! This way, you link your project to yourself, tell a personal story, and show that your path to valid results was not easy. With this arc of suspense, your audience will listen with even more interest. You show you have overcome all the hurdles and challenges to conclude your project successfully.

Create an arc of suspense with surprising twists and emotional intensifiers:

- How do I get into my short story or anecdote? What are the surprising twists and unique, unexpected moments that will take my audience's breath away?
- Can I use props to add visuals or unexpected sounds to my Science Pitch?

13.6   Storytelling with Randy Olson's Narrative Spectrum

Can you tell your project story in a dramaturgically exciting way? Marine biologist, film producer, and author Randy Olson (2015) uses the "Narrative Spectrum" for science stories that can be easily transferred to and applied for business presentations.

Olson warns against simply listing content instead of telling it: since these are not stories, they are boring. On the other hand, if you include stories with too many twists and turns, you will irritate your audience. The most intriguing narratives focus on two or three central topics. They contain a surprising twist and a short conclusion. An ideal basis for an entertaining presentation!

Here is a shortened example of a presentation from one of my earlier research projects, my doctoral thesis on a sequence of soils formed on marine terraces in the coastal area of Metaponto in southern Italy:

1st example, without narrative: "Here I present the results of the soil carbonates ... [details] **AND** the iron oxides show ... [details] **AND** the soil micromorphology shows ... [details] **AND** I would also like to present the results ... [more details]."

The mere enumeration gets lost in detail and is not put into context. The result is boring content without connection, therefore without a real narrative.

> *2nd example, with too much narrative:* "We see a gradually increasing soil reddening slightly correlating with their age … [details]. **DESPITE** that, we have also surprisingly found increased carbonate levels … [details]. **HOWEVER**, the results of the iron oxides do not match this because ... [details]. **YET**, we recognize from the micromorphology that there is shrinkage and swelling in the soil ... [details]."

Such endless loops only confuse our audience unnecessarily. They are easily distracted, and the contact between the presenter and the audience is lost.

However, the following narrative sounds much more interesting:

> *3rd example, optimal narrative:* "Let's talk about soils that formed before the last ice age. They were formed on marine terraces near the sea around Metaponto in southern Italy. Parallel to their age, the soils' redness **AND** weathering degree increase [details]. The distribution of iron oxides also reflects this, as we can see here [more details]. **BUT** surprisingly, the number of carbonates increases simultaneously [details]. Our micromorphological analyses reveal these as secondary carbonates [details]. We can, **THEREFORE**, conclude that the soils of older terraces were enriched or secondarily calcified by wind-blown sediments from North Africa during dry climate periods. For climate research based on soils, several time courses can occur parallel in soil chronosequences and must be considered individually. Only in this way is a comprehensive view possible ... [conclusion and outlook]."

To summarize:

- *AND* combines results and puts them into context. The big picture becomes visible.
- *BUT* creates new moments of tension. The narrative direction changes because the usual path is left, and something new and interesting comes into play. There is a lot of potential here for emotionally charged narratives!
- *THEREFORE* tells us about the consequences. The conclusions become clear, highlighting new knowledge and/or behaviors. This can lead to follow-up projects.

As is often the case, an optimal narrative lies in the middle: a captivating story with a certain complexity encourages the audience to think along without confusing them.

You can also replace "and—but—therefore" with synonyms. This makes your narrative style even more variable and personal. You can also use this structure several times in your presentation: if you present three project results, you can use "and—but—therefore" or its synonyms three times in a row.

The longer your presentation lasts, the more I recommend you use this pattern flexibly. Don't stick to it too rigidly, but use it as the basis for a varied, exciting narrative of your project.

Now Summarize the Story of Your Project in Just Under Three Minutes!

_____ and _____.

But _____.

Therefore _____.

13.7   Storytelling with Emma Coats' Pixar Pitch

The Pixar Pitch is pretty much an extension of Olson's Narrative Spectrum. It adds even more variety to your story. In the Pixar Pitch, you briefly summarize your story in a few paragraphs: address the initial situation and the change, explain why the change occurred, and briefly describe the new situation.

Your Pixar Pitch forms the basis for more detailed stories. This could be your 3-min Science Pitch for a more extended conference presentation, a publication, or introducing yourself at a networking event.

Emma Coats (2011) revealed the code of the narrative story during her time as a story artist at Pixar Animation Studios. The story structure of Nemo is a very good basis for developing your own stories. It continues to be the basis for all films from the Pixar film studios (Table 13.4). It is about six film sequences that can be found in every film, illustrated here using the story of "Finding Nemo" as an example. The Pixar Pitch template can be easily transferred to project narratives. My PhD thesis (Wagner 2009) provides exemplifies this (Table 13.4).

Adapt the wording to your style. Instead of "Once upon a time …", state the initial situation. You can replace "Because of that …" with terms such as "therefore", "hence", or "consequently". You can also add more details to the six sections to optimize your speaking time. The structure of the Pixar Pitch ensures a coherent dramaturgy in your presentation.

13.8   Tell the Story Behind Your Data, Facts and Figures

Your data, facts, and figures should always be recorded precisely. If your methodology is suitable, your numbers will also be accurate and provide a neutral statement for the respective time. Let's assume you present some of your research results using statistical graphics. The graphs should be easy to grasp. You build up a rather complex graphic step by step so your audience can always follow you. This is a challenge, especially for short presentations!

Table 13.4 Structure of the Pixar Pitch with examples of "Nemo" and a PhD thesis

Pixar pitch structure	Finding nemo (Coats 2011)	PhD thesis (Wagner 2009)
1. Once upon a time there was …	… Marlin, a widowed fish who was extremely protective of his only son, Nemo.	… soils in the Metaponto region in southern Italy. They were formed on marine terraces before the last ice age.
2. Every day …	… Marlin warned Nemo of the ocean's dangers and implored him not to swim far away.	… both the reddening and the degree of weathering of the soils have increased more or less continuously. The distribution of iron oxides also reflects this.
3. One day …	… in an act of defiance, Nemo ignores these warnings and swims into the open water.	… carbonates enriched these soils.
4. Because of that …	… he is captured by a diver and ends up in the fish tank of a dentist in Sydney.	… we also carried out micromorphological analyses to learn more about the specific properties of these soils and sediments.
5. Because of that …	… Marlin sets off on a journey to recover Nemo, enlisting the help of other sea creatures along the way.	… we can prove that the carbonates of the soils were not completely formed in situ but were partly introduced from outside.
6. Until finally …	… Marlin and Nemo find each other, reunite, and learn that love depends on trust.	… we can conclude that the soils on older marine terraces were secondarily calcified by wind-blown sediments from North Africa during dry periods. They are composed of a mixture of locally formed soils and externally introduced sediments.

To turn it into a story, you can add context to your data: how and under which conditions did you collect your samples or data? Are there social, economic, or environmental aspects in the background that you should be aware of?

Or you can talk about the changes and recent developments illustrated by your data. Again, the visual presentation should be as simple as possible to be captured within a few seconds. This often leads to a conflict between the scientific requirement to stay precise and the public's requirement to grasp information quickly and easily. We should differentiate between publications and presentations. In publications, precision is essential so that results can be reproduced. In presentations with a limited time horizon, simplification, and quick understanding gain priority over precision. A prominent example is the charts in the news about voter migration (Gantenberg 2023): in a daily or weekly newspaper, the reader has time to take a closer look at the chart. In a talk, the limited time overwhelms many in the audience.

The most suitable way to translate data into a narrative is to use concrete and sometimes striking examples from real life:

The 2002 federal election campaign in Germany was characterized by a story that is said to have contributed a decisive advantage for the later Federal Chancellor Gerhard Schröder. His opponent, Edmund Stoiber, could recite all the statistics on unemployment and employment rates in detail. Schröder, on the other hand, recounted the personal story of a man who had been unemployed for a long time and vividly described his family situation. He was narrowly re-elected as Chancellor. We want to hear stories and put ourselves in someone else's shoes.

Another option is translating methodical procedures into a generally understandable and visual language. Global Keynote Speaker Michael Wader (2024) shared two brief examples in a LinkedIn discussion of how to explain methodological approaches to our audience: "[T]rying to track microscopic particles by looking through a microscope is like trying to catch a goldfish swimming in a fish tank with a spoon.", and "trying to find a single defective cell in a sample of blood is like trying to find one specific person going to work in New York City from 3,000 meters above the city in an airplane."

Our audience is much more likely to identify with stories than with a list of statistically and scientifically correct data. But do not get me wrong: for science to remain reliable, comprehensible, and reproducible, precise information on methodology, procedure, and measurement is essential. Publications in renowned scientific journals are suitable for this. For presentations, a certain degree of fuzziness is usually necessary to ensure that we remain understandable and reproducible (Fig. 13.1). Especially for short presentations, you cannot avoid making the content more concise and entertaining.

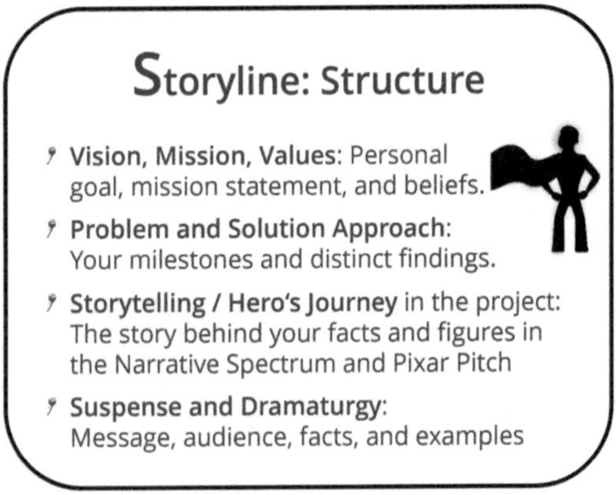

Fig. 13.1 Key Takeaway: In your storyline, well-founded scientific content becomes immediately understandable (source: author illustration)

References

Asmus, Frank 2021. Impact! Wie Sie sich und andere überzeugen – The Power of Influence. 296 pages, Goldegg.

Browning, Jo Filshie 2021. Scientifically Speaking. 168 pages, Practical Inspiration Publishing.

Campbell, Joseph 1949. The Hero with a Thousand Faces. Pantheon Books. 1st edition, Bollingen Foundation, 1949. 2nd edition, Princeton University Press. 3rd edition, New World Library, 2008.

Coats, Emma 2011. Pixar Story Rules. Source: Price, David A. Stories from the Frontiers of Knowledge. Pixar Story Rules 2011. https://www.davidaprice.com/pixar-story-rules. Accessed 13 Jan 2024.

DAAD – Deutscher Akademischer Austauschdienst 2024. Motto. https://www.daad.de/en/the-daad/who-we-are/motto/. Accessed 11 Feb 2024.

DFG – Deutsche Forschungsgemeinschaft 1955, 2023. Statutes of the Deutsche Forschungsgemeinschaft. https://www.dfg.de/resource/blob/17 5686/0cc1f65369941280c058204e656b47e0/dfg-satzung-en-data.pdf. Accessed 29 Jan 2024.

Dr. Rainer Wild-Stiftung 2024. Unsere Vision und Mission. https://www.gesunde-ernaehrung.org/unsere-vision-und-mission.html. Accessed 14 Jan 2024.

Fraunhofer SCAI 2024. ManuBrain. Universell einsetzbare, skalierbare KI-Plattform für Industrielle Anwendungen. https://www.scai.fraunhofer.de/content/dam/scai/de/documents/Mediathek/Projektblaetter/ManuBrain_Projektblatt.pdf. Accessed 29 March 2024.

Gantenberg, Julia 2023. Eine Grafik sagt mehr als tausend Studien. Gastbeitrag Wissenschaftskommunikation.de, https://www.wissenschaftskommunikation.de/eine-grafik-sagt-mehr-als-tausend-studien-72991/. Accessed 13 Jan 2024.

Google 2024a. About Google. https://about.google/intl/ALL_us/. Accessed 4 Feb 2024.

Google 2024b. Ten things we know to be true. https://about.google/philosophy/. Accessed 04 Feb 2024.

Google Bard 2024. What is Google's vision? https://bard.google.com/chat/dedacef780eabaa5. Accessed 29 Jan 2024.

Kennedy, John F. 1962. John F. Kennedy Podium. Space Center Houston. https://web.archive.org/web/20211026110739/https://spacecenter.org/exhibits-and-experiences/starship-gallery/kennedy-podium/. Accessed 29 Jan 2024.

LinkedIn Careers 2024. Culture & Values. https://careers.linkedin.com/culture-and-values. Accessed 29 Jan 2024.

Martins, Julia 2023. 5 Tipps zur Formulierung überzeugender Unternehmenswerte, die Ihre einzigartige Unternehmenskultur widerspiegeln (mit Beispielen). https://asana.com/de/resources/company-values-examples. Accessed 29 Jan 2024.

Netflix 2024. Netflix Culture – Seeking Excellence. https://jobs.netflix.com/culture. Accessed 29 Jan 2024.

Olson, Randy 2015. Houston, We Have a Narrative. Why Science Needs Story. University of Chicago Press, 256 pages (Paperback).

Sinek, Simon 2009. How great leaders inspire action. TEDx Puget Sound. https://www.ted.com/talks/simon_sinek_how_great_leaders_inspire_action. Accessed 14 Jan 2024.

Sinek, Simon 2011. Start With Why. How great Leaders inspire everyone to take action. 256 pages, Penguin LLC US.

Vogler, Christopher 1992. The Writer's Journey: Mythic Structure for Storytellers and Screenwriters. 297 pages, Image Book Company.

Wader, Michael 2024. Comment on the LinkedIn post "Are Scientists Competent Communicators Presenting Their Research?" by Dr. Stephen Wagner. https://www.linkedin.com/feed/update/urn:li:activity:7168122667449970688?commentUrn=urn%3Ali%3Acomment%3A%28activity%3A7168122667449970688%2C7168138010079780864%29&dashCommentUrn=urn%3Ali%3Afsd_comment%3A%287168138010079780864%2Curn%3Ali%3Aactivity%3A7168122667449970688%29. Access 28 Sept 2024.

Wagner, Stephen 2009. Soil (Chrono-) Sequences on Marine Terraces. Pedogenesis in two coastal areas of Basilicata and Agrigent, Southern Italy. 327 Seiten, Hohenheimer Bodenkundliche Hefte, Heft 93, PhD Thesis. https://hohpublica.uni-hohenheim.de/bitstreams/7c965f5a-a633-413b-89a7-92170d9d72fa/download. Accessed 16 Nov 2024.

Wagner, Stephen 2023. Public Speaking in Cameroon. https://redeland-schaften.de/en/public-speaking-in-cameroon/. Accessed 14 Jan 2024.

Performance: Your Compelling Stage Presence with Character

14

Abstract

Effective communication is based on a targeted stage presence and is one of the building blocks for a convincing performance. So, think carefully about the setting for your appearance. Authenticity matters more than perfection. You will be particularly authentic and approachable if you share personal experiences and enthusiasm for the project. Illustrate special moments of your project vividly and with many facets.

If your presentation is well-founded in terms of content and expertise and adds value for your audience, anecdotes contribute to an authentic and convincing performance. A personal appearance almost automatically ensures a melodic, dynamic voice and lively facial expressions and gestures, unlike artificially trained behavior. In doing so, you reveal a significant part of your personality to your audience. A congruent, authentic stage presence demonstrates pathos and comes across as credible.

For a convincing performance, think in advance about how to stage your presentation skillfully. Get familiar with the stage in advance. You may want to present your Science Pitch as if you were talking to your best friends and colleagues. Authenticity is more important than a perfectly polished presentation you recite as if you have memorized it. Think about your emotional connection to your audience. Show your human side by sharing your motivation, passion for the topic, and ESPRIT. You can highlight special moments for this: combine professional expertise with your personality. That way, you are much more likely to be remembered by your audience than in an average, rattled-off presentation.

14.1   Stage Yourself: Your Stage Presence Is Your Gamechanger

A well-planned script will ensure that you reach your audience. In a short Science Pitch, you do not need to discuss speech structure; keep the theoretical part as brief as possible. Personal stories are much more effective: link your practical experience to the problem or current challenge. Then, outline the path to your solution. This provides the basic framework for your Science Pitch.

Effective communication thrives on the stage and amplifies your content.

Clarify the context of your Science Pitch for yourself:

- How do I dress?
- When will I be on stage?
- Is my stage small, or will I use a large theater stage?
- How close will I get to my audience?
- Will I speak freely, with a microphone or with a headset?
- Do I work with meaningful slides?
- Will I show props, sounds, a video, or a prototype of my model?
- Which special moments and surprising twists do I emphasize?
- Do I refer to previous or subsequent speakers with a related topic?

Consider the setting for your presentation in detail. It is best to test your content and messages extensively and in advance on the original stage. The better prepared you are, the more likely you are to show your presence at the crucial moment: you are entering into a relationship with your audience. You are not just delivering an isolated presentation.

14.2 Authenticity Counts More Than Perfection

A compelling talk depends on your emotional connection to your audience. Can you inspire them? Do you get them to act, to do something practical? Here, your vision and mission (see Sects. 13.1 and 13.2) come into play. The values (see Sect. 13.4) also play a role here: they show what you currently stand for with your project and where you want to develop. Do you act accordingly? Do you translate words into action?

There is a saying, "Walk your Talk". This is about congruence, the match between what you say and what you do. Do your content and body language match? Your audience has fine sensors and will quickly notice whether everything fits. Avoid claiming the experiences of others as your own. Instead of telling other people's stories, bring in first-hand experience.

How do you show yourself as authentic and approachable without being insincere?

> Answer these questions for your stage performance:
>
> - How did my project shape my personality? How did I contribute to the project? How much do I identify with it?
> - Do I present myself as authentic and approachable without being insincere?
> - How do I convince with my performance?
> - Why am I/is my team particularly suited to implement the presented project?

You appear more authentic and approachable when sharing personal experiences and expressing emotions. Your voice will sound much more melodic; your facial expressions and gestures are congruent and do not appear imposed. Are you self-confident? Then you have a great chance to perform according to your values. You do not need specific body language training for this, either.

To deliver an authentic and competent performance, you need to bring well-established communication skills and a stable level of self-confidence. This is why regular training and stage experience are essential for a successful presentation. According to Browning (2023), science "must be communicated well" to "have [a] true impact". Yet many scientists are "worried that their work might be miscommunicated, misinterpreted, or misunderstood. But in the attention economy, this is exactly why communication skills are essential. Nothing is more inspiring than seeing a scientist communicate their work confidently and effectively." Browning concludes that "science has the impact it deserves, and the people who need to know can fully understand it."

Nevertheless, although your performance should be high, it does not need to be perfect. Minor mistakes show your human side and are forgiven by the audience.

14.3   Expertise and Passion: A Powerful Team

Combine your professional expertise with personal passion. This will make it much easier for you to convince your audience.

So, answer the following questions:

- What do I love about my research?
- What motivates me in particular? Why did I choose my topic or project?
- What moments of surprise can I incorporate into my talk?

My specialty at the time was paleopedology: as a geographer, I spent many years researching soils formed during or before the last ice age. I was thrilled that we could recognize and compare regional and global developments in recent Earth history by studying soils. As a haptic and visual person, I love the practical work in various landscapes: from the coast to savannahs to arid regions, we gain insights into spatial relationships. I could touch, feel, and analyze soils with my hands. I was also fascinated by color changes in soils, especially reddening degrees. This brief insight shows a lot of potential for personal anecdotes and stories. I still incorporate these into my speeches nowadays.

Condense your special moments into stories and anecdotes that you can recount in a nutshell. This can be a real "earcatcher" (Richter and Münzner 2020) by being concrete and specific, possibly enriched with surprising facts or curiosities. Anecdotes that match the content show your attitude to the topic and what excites you about your research. This way, you become visible as the person behind the project. It is a valuable component of your Science Pitch.

Consider specific personal and professional experiences that could add value for your audience. For inspiration, I suggest Deepika Kurup's TEDWomen Talk (2016).

Deepika Kurup speaks as a young scientist. At 14, she already had a grand vision of solving the global water crisis. This motivated her to develop a cost-effective and environmentally friendly water treatment method. In her TEDWomen Talk, she combines Science Pitch and personal storytelling. Kurup introduces the topic within the first 3½ min: she reveals her motivation through short stories and personal photos and presents her current project simply and understandably. She radiates enthusiasm and passion and captures her audience from the very first moment.

Write down special moments from your project, studies, or professional life.

Describe them vividly in all their facets:

- What do I love about my research? What interests, inspires, or fascinates me in particular? In which moments do I passionately pursue my project (Fig. 14.1)?
- Special moments in my project: where and when did I experience something extraordinary? How does it make a difference for me and my project?
- How do I translate these moments into visual, vivid statements? How do these fit into my Science Pitch?
- When I talk to colleagues and friends about my project, at which points do they react with particular interest or ask questions?

14.4   Enthusiasm Adds Momentum to Your Voice and Body Language

Besides your preparation and routine, your attitude and your mindset are also important. If you are enthusiastic about your topic and have sufficient self-confidence, you will radiate this naturally to your audience. On the other hand, your audience will perceive artificially trained facial expressions and gestures as inappropriate and, therefore, untrustworthy.

I emphasize that you share your enthusiasm for your project by anecdotally incorporating special moments. If you are enthusiastic, your eyes will light up, your gestures will become more open, and you will turn to your audience with open body language.

In addition to numerous talks with high-quality feedback, I have gained self-confidence over the decades because I am convinced of my expertise and performance and am well-prepared for content. This has not always been the case (Wagner 2020).

Authenticity is important for any speech. Present yourself as you are. Your voice will be more dynamic and melodic if you do not rattle off your script but emphasize special, inspiring moments. If you are enthusiastic, you are much more likely to use your arms and hands without training yourself. Your audience will notice that.

Authenticity leads to lively body language and a melodic voice. An intriguing study by behavioral scientist and body language trainer Van Edwards (2015) shows hand gestures and smiling as particularly effective tools that captivate an audience. Van Edwards correlated the number of hand gestures and smiles in presentations with the number of online views of TED talks. Authentic body language creates credibility. Van Edwards also revealed that smiling makes speakers appear smarter and that intelligence is sometimes rated higher. The smile should not be artificial or forced but genuine.

Vocal volume and tempo, vocal melody, emphasis, and pauses add character to your speech and contribute to a dynamic presentation. Your voice conveys a significant part of your personality to the audience. You can warm up your voice with simple training. Julian Treasure, an expert in sound and communication, provides short exercises in his TED Talk (Table 14.1):

Repeat each exercise as often as you like, but spend at least 30 s on each round.

When testing your presentation, I recommend recording and evaluating it using a smartphone camera. Then, listen to your audio without watching the video: assess what you said that was good and what you would like to improve. Afterward, watch the video without sound to recognize your charisma and enthusiasm, or adjust your stage presence if you want to present even more dynamically. Only at the end should you watch the entire video with sound.

You can use these insights to revise your Science Pitch. Working on just one or two specific points before rehearsing again is mostly sufficient. Always stay aware of your strengths: self-praise is right!

Table 14.1 Warm up your voice with an intense exercise (Treasure 2013)

Step	Exercise
1.	Arms up, deep breath in, and sigh out, "Ahhhhh", like that. One more time. "Ahhhh", very good.
2.	Now we're going to warm up our lips, and we're going to go "ba, ba, ba, ba, ba, ba, ba, ba". Very good.
3.	And now, "brrrrrrrrrr", just like when you were a kid. "Brrrr". Now your lips should be coming alive.
4.	We're going to do the tongue next with exaggerated "la, la, la, la, la, la, la, la, la". Beautiful. You're getting really good at this.
5.	And then, roll an R. "Rrrrrrr". That's like champagne for the tongue.
6.	Finally, and if I can only do one, the pros call this the siren. It's really good. It starts with "we", and goes to "aw". The "we" is high, the "aw" is low. So you go, "weeeaawww, weeeaawww."

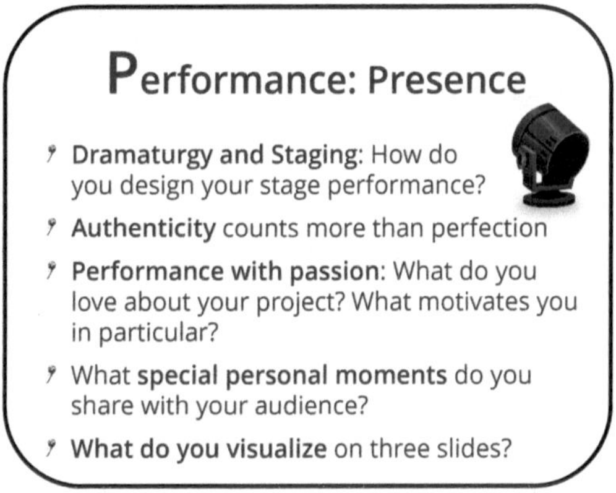

Fig. 14.1 Key Takeaway: A successful performance requires stage presence and passion (source: author illustration)

References

Browning, Jo Filshie 2023. Science Communications matters and how to do it better. TEDx Basel. https://www.youtube.com/watch?v=7Rt8sgt7gNE. Accessed 31 Jan 2024.

Kurup, Deepika 2016. A young scientist's quest for clean water. TEDWomen 2016. https://www.ted.com/talks/deepika_kurup_a_young_scientist_s_quest_for_clean_water. Accessed 14 Jan 2024.

Richter, Kay-Sölve and Münzner, Richard 2020. Viel mehr als nur Körpersprache – Executive Presence. Wie Sie als Führungskraft überzeugend auftreten, wenn es darauf ankommt. GABAL Verlag GmbH Offenbach, 240 pages.

Treasure, Julian 2013. How to speak so that people want to listen. TEDGlobal 2013. https://www.ted.com/talks/julian_treasure_how_to_speak_so_that_people_want_to_listen. Accessed 14 Jan 2024.

Van Edwards, Vanessa 2015. 5 Secrets of a Successful TED Talk. https://www.scienceofpeople.com/secrets-of-a-successful-ted-talk/, page view on January 14, 2024. From: Prato, Alison 2015. Does body language help a TED Talk go viral? 5 nonverbal patterns from blockbuster talks. https://blog.ted.com/body-language-survey-points-to-5-nonverbal-features-that-make-ted-talks-take-off/. Accessed 14 Jan 2024.

Wagner, Stephen 2020. Develop Self-Confidence and Become a Convincing Speaker. https://redelandschaften.de/en/self-confidence-and-convincing-as-speaker/. Accessed 14 Jan 2024.

Relevance: How Significant Is Your Project for Others?

15

Abstract

Research projects aim to provide new scientific insights. You should also emphasize the knowledge transfer into practice and the social benefits and explain for whom your project is particularly interesting and why. This involves the latest findings, the realistic potential for implementation, and a corresponding roadmap for your ideas. With whom do you enter an ongoing dialog outside of your project? It will quickly reveal the innovative solutions that could enrich science, business, and society.

The goal of research projects is to gain new scientific knowledge. In basic research, you will explain the knowledge transfer into practice. The German Research Foundation (DFG) describes it on its website as "the scientific findings and results generated by basic research" (DFG 2024). More particular, it is about the innovative nature of your research approach: have you developed a model that depicts future trends? Do you guide on controversial issues?

15.1   Your Target Group: Who Is Interested in Your Project—and Why?

The answer to this overarching question is an important pillar of your Science Pitch. The art of your brief presentation lies in distilling the essence of your project into a few sentences within the shortest possible time.

Consider these specific questions:

- Can I explain the relevance and future potential of my research topic?
- Can I depict future long-term trends? Which ones?
- What is the practical relevance of my idea in my field of research and beyond?
- Which industries, organizations, and companies utilize the knowledge I generate?
- Can I translate my findings into specific recommendations and implement them in practice?
- In the case of a methodological or procedural focus; who will apply my methods or procedures in practice?
- Who reads my publications? Who else might they be of interest to?

In your Science Pitch, you present a specific situation, such as a practical example or personal experience. This is how you shape future reality and identity—and identity creates identification.

15.2   Feasibility: Can You Implement Your Project Idea?

To reveal the implementation potential of your ideas, you need a realistic roadmap. Make sure that the significance of your ideas becomes immediately clear. You are not just providing new insights; they must also be realistic and implementable.

The following questions are important for the feasibility of your project:

- Is the funding for my project based on solid ground?
- Are there legal hurdles to overcome to implement my project?
- Can I realistically implement my project within the given time frame?
- What space and room requirements do I have?
- Do I need laboratory materials and technical equipment?
- Do I rely on the help of technical and student assistants?

As in every section of your Science Pitch, it is also important to highlight only the most important aspects for your target audience (Fig. 15.1).

15.3   Can You Scale Your Project?

Can you reproduce and automate your innovation beyond the project with minimal additional effort? Digitization, effective marketing, sales, and expansion into neighboring disciplines and other markets open new opportunities for you. From an economic perspective, it is all about the increased revenue generated by your idea.

Do you communicate the societal benefits and relevance of your project? How do you convey the results of your research to different target groups? In addition to publications, lectures, and colloquia, this can include excursions, exhibitions, panel discussions, workshops, interviews, videos, or podcasts. Seek continuous dialog with the outside world and actively engage in debates.

In a broader sense, this includes the exchange between basic and applied research, on the one hand, and industry and business, on the other. In this way, your project contributes to innovative solutions in the technology sector. Furthermore, scientific consulting has an impact on politics, business, society, and culture.

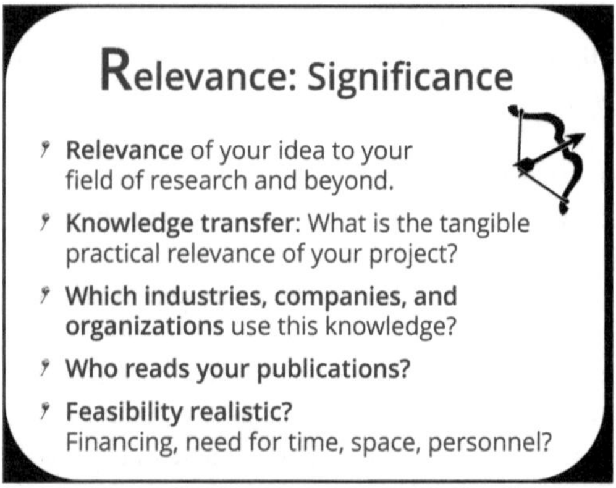

Relevance: Significance

- *Relevance* of your idea to your field of research and beyond.
- *Knowledge transfer*: What is the tangible practical relevance of your project?
- *Which industries, companies, and organizations* use this knowledge?
- *Who reads your publications?*
- *Feasibility realistic?* Financing, need for time, space, personnel?

Fig. 15.1 Key Takeaway: You should highlight the relevance of your project to the outside world (source: author illustration)

Reference

DFG – Deutsche Forschungsgemeinschaft 2024. Funding Initiatives – Knowledge Transfer. https://www.dfg.de/en/research-funding/funding-initiative/knowledge-transfer. Accessed 31 Jan 2024.

Innovation: Highlight Your Unique Selling Proposition (USP)

16

Abstract

You should emphasize your project's pioneering factor and innovative content, highlighting unique features and setting yourself apart from others. You must address both the project's special added value and specific benefits, as well as its potential economic output. This is about originality and the relevance of aspects that have received limited attention so far. It promises results that reach beyond existing knowledge.

Science and research thrive on discoveries and new insights. Science creates new knowledge. It lays the foundation for integrating new things into our everyday lives, thus advancing humanity. Some discoveries may happen by chance; most projects are pursued through a clear vision with pioneers on a mission. They are intrinsically motivated and live out their creativity. Bringing all these aspects together leads to a unique selling proposition (USP). This section examines the facets up to the wording of your USP for the Science Pitch.

© The Author(s), under exclusive license to Springer Fachmedien 109
Wiesbaden GmbH, part of Springer Nature 2024
S. Wagner, *Science Pitch*,
https://doi.org/10.1007/978-3-658-44844-8_16

16.1   Pioneer Factor: Your Innovation Broadens Horizons

As a scientist and researcher, you will most likely contribute new ideas. Research thrives on innovative content! That is why it is so important for any kind of communication. In events like the "Falling Walls Pitches" (Falling Walls Lab 2011–2024), the innovation or pioneering factor is even of paramount importance if you want to succeed while competing with other researchers. You should also illustrate the innovative character of your project (Fig. 16.1).

The USP emphasizes the additional benefit, and the special added value of the project for science, economy, and society.

> Answer the following questions before you present your Science Pitch:
>
> - How does my project stand out from others? Do I emphasize the subtle differences compared to other projects?
> - How innovative, original, unique, and groundbreaking is my project?
> - What future potential does my project have? What specific benefits and added value does it bring—and for whom?
> - More specifically: do I highlight the benefit of my project for a particular market/research segment? In which branch? Do I address the public at large?

Science thrives on discoveries. You can highlight these by comparing them with previous projects and their results, much like in a competitive analysis. From an economic perspective, innovations also involve the additional revenue you can generate through a patent or newly developed product.

16.2 Originality and Creativity: The USP of Your Project

Is your research topic new and relevant? Is it currently overlooked or barely addressed? This might be your opportunity to achieve high originality!

Originality is defined as "the degree to which a scientific discovery provides subsequent studies with unique knowledge that is not available from previous studies" (Shibayama and Wang 2019, pp. 409–410). In other words: originality is based on unexpected, surprising results that reach beyond existing knowledge. In terms of scientific relevance, originality demonstrates the degree of creativity that you can distinguish from other research achievements. According to Polanyi (1969, quoted by Heinze 2013, p. 928), originality is also based on surprise: "Contributions that obtain unexpected and thus surprising results are usually classified as original." Ideally, originality should be connected with scientific relevance (Heinze 2013). This way they will stand out from the mainstream, are not controversial, and open space for new questions and topics. They elevate our knowledge to a new level.

On the other hand, creativity is defined as a generation of novel and useful ideas. Creativity in research results in practical values or significance that have not been developed or applied before. According to Simonton (2012), creativity requires novelty, utility, and surprise as indispensable factors, while Cropley (2020) highlights novelty, relevance, and effectiveness, but also morality and ethicality as indispensable factors for creativity.

It is equally about your creativity and independent thinking, i.e. the substantially new and unique approach to your project.

This particularly includes original research with new scientific insights:

- New data
- New interpretations of existing data
- Advanced or entirely new methods
- New solutions to previously unsolved problems

Make sure to distinguish your project from the already existing knowledge. Table 16.1 highlights summarized examples of USP's from scientific projects:

Table 16.1 Examples of unique selling propositions (USP's) for science projects

Project	Unique Selling Proposition (USP)
Center on Gender Equity and Health (GEH), University of San Diego (McDougal et al. 2018)	Result: child marriage is promoted not only by extreme poverty or lack of education but also by decision-makers, parent-child relationships, social norms, and place of residence.
Pilot project at Tübingen University Hospital (Alzheimer's Research in Dialog 2021)	Exploring new therapeutic approaches between microglial cells (immune cells of the brain) and Alzheimer's disease. Here: a unique human model for studying microglial cells.
Institute of Soil Science and Land Evaluation, University of Hohenheim (GEPRIS 2024)	Pedo-dating of marine terraces of unknown age via soil chronofunctions. Space-time reconstruction of Quaternary littoral displacement in southern Italy.
2012 Discovery Education 3M Young Scientist Award (Discovery Education 2012; Patel 2013)	Cost-effective method for water purification: sunlight and a photocatalytic composite of TiO_2 and $AgNO_3$. Coliform and E-coli contamination was almost fully eliminated.

In addition, Table 16.2 picks up examples of USP's for companies:

Table 16.2 Unique selling propositions (USP's) for companies

Company	Unique Selling Proposition (USP)
Canva (2024)	Empowering the world to design.
Colgate (Lorincz 2024)	Improve mouth health in 2 weeks.
IKEA (2024)	A world of inspiration for your home.
M&M's (Lorincz 2024)	Melt in your mouth, not in your hand.
The North Face (2024)	Shaping the future of human/nature.

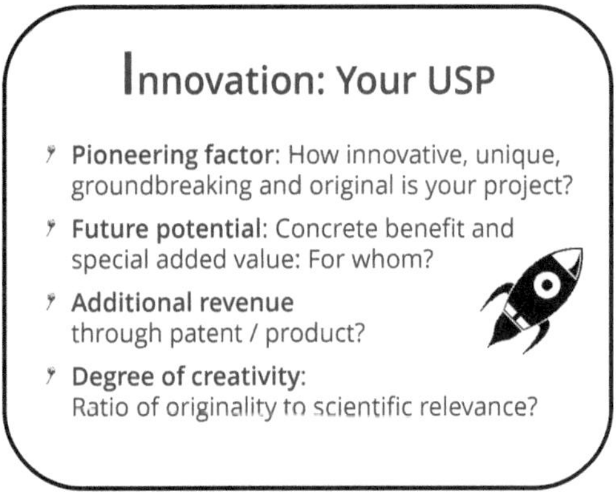

Innovation: Your USP

- ⚡ **Pioneering factor**: How innovative, unique, groundbreaking and original is your project?
- ⚡ **Future potential**: Concrete benefit and special added value: For whom?
- ⚡ **Additional revenue** through patent / product?
- ⚡ **Degree of creativity**: Ratio of originality to scientific relevance?

Fig. 16.1 Key Takeaway: Illustrate the innovative character of your project (source: author illustration)

References

Alzheimer-Forschung im Dialog 2021. Den Reinigungsdienst des Gehirns erforschen. Dr. Gaye Tanriöver mit dem Helga Steinle-Preis ausgezeichnet. Nr. 13, Frühjahr 2021. https://www.alzheimer-forschung.de/forschung/forschungsprojekte/projektdatenbank/projekt/einzigartiges-menschliches-modell-zur-untersuchung-von-immunzellen-im-gehirn/, Accessed 14 Jan 2024.

Canva 2024. Empowering the world to design. https://www.canva.com/about/. Accessed 10 Feb 2024.

Cropley, Arthur J. 2020. Definitions of Creativity. In Runco, Mark A. and Pritzker, Steven R. (Editors.), Encyclopedia of Creativity, 3rd edition, 315-322. San Diego, CA: Academic Press. https://doi.org/10.1016/B978-0-12-809324-5.23524-4

Discovery Education 2012. Discovery Education and 3M announce 2012 Science Competition Winner. https://www.discoveryeducation.com/details/discovery-education-and-3m-announce-2012-science-competition-winner/. Accessed 14 Jan 2024.

Falling Walls Lab 2011–2024. The global interdisciplinary platform for students and early-career professionals. https://falling-walls.com/lab/. Accessed 14 Jan 2024.

GEPRIS, Geförderte Projekte der Deutschen Forschungsgemeinschaft 2024. Sauer, Daniela. Ableitung von Bodenchronofunktionen auf Meeresterrassen in Süditalien zur Pedo-Datierung. DFG-Projektnummer

41836859. https://gepris.dfg.de/gepris/projekt/41836859. Accessed 14 Jan 2024.

Heinze, Thomas 2013. Creative accomplishments in science: definition, theoretical considerations, examples from science history, and bibliometric findings. Scientometrics 95, 927-940. https://link.springer.com/article/10.1007/s11192-012-0848-9

IKEA 2024. IKEA Global. https://www.ikea.com/global/en/. Accessed 10 Feb 2024.

Lorincz, Nikolett 2024. Marketing: What is a Unique Selling Proposition? With 15 USP Examples (2024). https://www.optimonk.com/what-is-your-unique-selling-proposition-usp-examples/. Accessed 10 Feb 2024.

McDougal, Lotus, Jackson, Emma C., McClendon, Katherine A., Belayneh, Yemeserach, Sinha, Anand and Raj, Anita 2018. Beyond the statistic: exploring the process of early marriage decision-making using qualitative findings from Ethiopia and India. BMC Women's Health, 18/2018: 144, 16 pages. https://bmcwomenshealth.biomedcentral.com/articles/10.1186/s12905-018-0631-z. Accessed 14 Jan 2024.

Patel, Nita 2013. America's Top Young Scientist: Kurup Scores Top Prize [Pipelining: Attractive Programs for Women]," in IEEE Women in Engineering Magazine, Vol. 7, no. 2, pages 37-39. https://doi.org/10.1109/MWIE.2013.2280411. https://ieeexplore.ieee.org/document/6661485/authors#authors. Accessed 14 Jan 2024.

Polanyi, Michael 1969. Knowing and Being: Essays by Michael Polanyi. 264 pages, Chicago University Press.

Shibayama, Sotaro and Wang, Jian 2019. Measuring originality in science. Scientometrics, 122, 409-427. doi: https://doi.org/10.1007/s11192-019-03263-0. https://link.springer.com/article/10.1007/s11192-019-03263-0

Simonton, Dean Keith 2012. Taking the U.S. Patent Office Criteria Seriously: A Quantitative Three-Criterion Creativity Definition and Its Implications. Creativity Research Journal 24, 97-106. https://doi.org/10.1080/10400419.2012.676974

The North Face 2024. Shaping the future of human/nature. https://www.thenorthface.eu/da_dk/about-us.html. Accessed 10 Feb 2024.

Take Home Message: Your Headline in One Sentence

17

Abstract

An engaging and entertaining Science Pitch requires a clear arc of suspense. For this, define a short, concise, and memorable take home message to capture attention and interest. You can underline it with several catchy yet understandable key messages for each section of your presentation. Ideally, this strengthens the connection between your idea, your vision, and your audience. This is where science can learn from advertising. Your message does not have to be exact in a scientific sense; that is reserved for longer presentations and publications.

Your concise take home message runs like a red thread through your Science Pitch and corresponds to a short, memorable sentence. It can correspond to the title of your presentation or your Science Pitch. An additional subtitle can lead precisely to your topic. Allow sufficient time for brainstorming and avoid convoluted sentences at all costs. You can also use AI tools like ChatGPT for initial inspiration.

Then, develop a motivating key message for each section of your presentation that contributes to your take home message. Support these like Steve Jobs did, with convincing facts and coherent examples. You can find further inspiration from one sentence pitches in science, business, literature, and advertising. Your audience should be able to remember your take home message for a long time.

S. Wagner, *Science Pitch*,
https://doi.org/10.1007/978-3-658-44844-8_17

Short, concise messages capture attention and increase the interest of your audience. Launch an unusual speech title and a catchy slogan to ensure that your presentation is memorable. Enrich your take home message with several key messages in each section of your Science Pitch, setting multiple highlights.

According to Asmus (2024), the take home message "is the fundamental, overarching statement that embodies the essence of a brand, a campaign, or an organization. It can express the vision, the mission, the purpose, a central value, the USP, or simply the positioning. A take home message is the foundation on which all communication efforts are built. It is designed to last and reflect the identity or philosophy of the organization. A broad spectrum of target groups is usually addressed." As an orientation, the ideal hook for a clear take home message contains six words (Asmus 2024).

This is what science can learn from journalism: getting to the point with a message that arouses curiosity! In a recent TEDx talk, Browning (2023) highlighted how journalists grab attention: they start with the most important information in the headline, adding relevance and audience interest. To let your audience understand "why science matters", Browning recommends focusing "on what science can do rather than how the science was done." Especially in a non-science community, scientists should talk about everyday practical outcomes instead of detailed theories with scientific terms.

17.1   Create Suspense in Your Audience's Minds

To present your Science Pitch in a convincing, engaging, and entertaining way, you need an arc of suspense. This will keep your audience's attention and is immensely important for any presentation.

According to director and top executive coach Frank Asmus (2021), we can build a substantial and compelling dramaturgy by answering the following questions about strategically skillful communication:

- What is the one take home message?
- What key messages will connect with and inspire my audience? Can I substantiate them?
- What expectations can I anticipate from my audience?
- What facts, examples, stories, and references will I include?

In your Science Pitch, you phrase your take home message in one short single sentence. In contrast, a key message is defined as a "bridge to certain target groups. They are more specific and serve to substantiate the take home message for certain target groups, campaigns, or contexts and "unlock" the audience for it. Key messages are more flexible and can be adapted to current events to maximize relevance and engagement with specific target groups. Key messages are the variable, often tactical tools at the message level" (Asmus 2024). You will find your key messages in the thematic section of your presentation.

Ensure your short presentation matches your audience's expectations: talk to them long before you deliver your Science Pitch, put yourself in their shoes, and find out what information is particularly interesting and valuable to them.

Substantiate your key messages with clear facts and evidence. Here, you can delve a little deeper into the details of your research project.

17.2   Less Is More: Why Short Messages Are so Successful

It is worth investing time in your one sentence pitches. Short, concise messages capture attention and increase your audience's interest. Through practice and testing, you will gain clarity and a good overview of what is essential in your Science Pitch: take at least 5 min, preferably 15 min, and write down all the variations that come to mind, no matter how unusual. In doing so, you summarize the extensive content of your project in a concise form. Short messages are substantial, meaningful, and easy to understand when they convey essential information. Ideally, your one sentence pitch creates a connection between your idea and vision on one hand and your audience on the other. Ideally, this will also match their interests and needs.

Advertising and marketing apply this principle by addressing the emotional level, directly appealing to our senses and imagination. Ideally, meaningful facts are underlined by your key messages. This encourages your audience to think along. While shortened content might be easily understood, it may also compromise scientific accuracy in the sense that you translate reliable and reproducible information into a non-academic style. Avoid overly complex sentences that your audience will not understand.

Imagine someone speaking to you in this way:

"For the presentation of scientific results, whether it is in a seminar, a colloquium, during the defense of a doctoral thesis, or as part of a conference presentation—as is also the case for any other scientific presentation such as colloquia—the scientist takes care of the use of elaborated sentences supported by numerous foreign-language technical terms, so that one expresses oneself correctly and precisely in every case, avoiding overly long sentences, whereby naturally also paying attention to clean language stylistics, accurate spelling, and precise punctuation."

> You can cut it down to a much shorter message: "You are presenting your project? Pay attention to your audience: use clean language!"

> Here is another example: "Perspectives for Climate Policy: The Potential Role of Climate Policy Agreements for the Millennium Development Goals and Realistic Transport Policy Measures to Mitigate and Slow Climate Change".

> You can summarize it in a short and crisp sentence: "Slow down for climate change!"

Here, you can apply the principle "Less is more!" Well-formulated messages are simple, concrete, and credible. You can use the information you have left out in the subsequent discussion or a personal conversation.

17.3   Your Take Home Message Arouses Your Audience's Curiosity

Ask yourself about the added value of your presentation. Condense it into a take home message with a short, memorable sentence, which may even be the title of your project. It runs like a red thread through your Science Pitch.

Sharpen your take home message to be so concise that you cannot leave anything out anymore. It becomes as memorable as Barack Obama's "Yes we can!" or Nike's "Just do it!" This is communication at its best since it gets to the point immediately. Take home messages correspond to one sentence pitches and are easily recognizable. There are countless examples in the news, as book titles and subtitles, and in social media. They leave out long-winded, detailed information and arouse the audience's curiosity.

You can use short metaphors to present content with a wink to contrast and compare content. When developing your one sentence pitch, consider your project's particular benefit. Brainstorm several one sentence pitches and later decide on the most suitable and original version.

Here are examples of one sentence pitches from a wide range of industries: from popular advertising and book titles to the positioning of small and medium-sized enterprises, and even science papers.

The list in Table 17.1 provides examples of particularly short take home messages.

Table 17.1 Examples of take home messages for advertising and business

Company	Take Home Message
Nike	Just do it
McDonald's	I'm lovin' it
Toyota	Let's go place
Apple	Think different
Coca Cola	It's the real thing!
Airbnb	Belong anywhere
Otis	Made to move you
Red Bull	Red Bull gives you wings
iPod, Apple (by Steve Jobs)	1000 songs in your pocket

The take home messages on the left side of Table 17.2 were developed by participants in my Science Pitch training during brainstorming sessions and the subsequent discussions—some of them translated from German. On the right-hand side of Table 17.2, you find take home messages, which I have been developing for my own business since 2015. I applied them mainly as slogans for websites, social media channels, business cards, and statements during presentations and workshops. I am currently using the slogans "I Take Science Out of the Blackbox" and "Present Your Research. Get to the Point".

In addition, you can also include your take home message as an attractive title for your PhD thesis or presentation, as in the following examples (some were translated from German):

Table 17.2 Take home messages: Ideas from pitch training sessions, and individual examples of the author

Ideas from Pitch trainings	Examples by Dr. Stephen Wagner
Heat? Pumps!	Create Knowledge Through Speaking.
Digitize your Rail!	I Take Science Out of the Blackbox.
Digitizing Bioprocesses.	The First (Last) Sentence is Yours!
Let's Start to Stop Animal Abuse.	Feedback for Your Presentation.
Networking Knows No Boundaries.	Presentations for Global Minds.
Construction Robotics? Yes, Please!	Less is More: Kill Your Darlings.
Better Automation Than Stagnation.	Science Touching Business.
Never Tired. Never Sick. Always Accurate.	Presentation is Fascination.
Steep Learning Curves Instead of High Hurdles.	Color Your Speech.
The Compound Interest Neck: Stairway to Success.	Think. Speak. Act.

- Evergreen or One-Hit Wonder? Career Paths of Musicians. (Pfeiffer 2018, 2024, translated)
- Police Information Systems: State of the Art and Current Developments. (Pfeiffer 2018, 2024, translated)
- Absolute and Non-Absolute Hearing: Factors Influencing the Recall of Keys. (Schlemmer 2006, translated)
- "You Find Yourself Standing on the Street and Stopping!": A User-Oriented Study on the Experience and Coping with Housing Shortage Among Women. (Baum 2020, translated)
- Reddening as climatic indicator? Investigations on Quaternary soils and soil sediments of the Balearic Islands. (Wagner et al. 2012).

In addition to your personality and authenticity, consider a clever dramaturgy of your performance.

Therefore, answer the following questions:

- If it were just one thing: what do I want my audience to remember?
- What is my unique take home message?

How do you organize your presentation? Your answer to this question is less trivial than it first appears. In most presentations, young scientists follow a conventional IMRAD-style slide format, added conclusions, a literature list, and a "Thank You for Your Attention" slide. But for presentations in general, and Science Pitches in particular, we should limit ourselves to essentials instead of copy-pasting publications.

Many aspire to present in a way that highlights their expertise and competence. At the same time, they often receive instructions from their professors and supervisors that a presentation must follow the above-mentioned format. From a presentation perspective, such a list does not add any value and does not arouse audience curiosity—on the contrary! In a Science Pitch, a slide showing the speech structure is redundant because it is common without revealing new insights and unnecessarily takes up valuable presentation time.

Long-winded sentences that contain almost everything the expert wants to present are equally pointless. What do you think of this poster title: "Decisive Aspects for the Sense of Belonging at the Workplace: Insights Gained from Comparing Permanent Employees with Freelance Workers. Presentation of the Results from Our Detailed Study"? (Table 17.3).

27 words, 59 syllables, and 163 characters later, I wonder how this title can be shortened to its essentials. Here are two suggestions: "Boosting Your Employees' Sense of Unity: Actionable Insights" and "Making Freelancers Feel as At Home as Your Staff—Study Reveals Latest Findings" (Table 17.3).

Table 17.3 Analytic comparison of long vs. short poster titles

Analysis	Original poster title	Analysis	Shorter poster title
27 words 59 syllables 163 characters	Decisive Aspects for the Sense of Belonging at the Workplace: Insights Gained from Comparing Permanent Employees with Freelance Workers. Presentation of the Results from Our Detailed Study.	8 words 17 syllables 53 characters	Boosting Your Employees' Sense of Unity: Actionable Insights.
		13 words 20 syllables 68 characters	Making Freelancers Feel as At Home as Your Staff—Study Reveals Latest Findings.

Both suggestions share a common structure: the first sentence is the main title, followed by an additional explanation. This approach breaks down the overly long original title. The titles are now shortened to 8 words, 17 syllables, and 53 characters, or 13 words, 20 syllables, and 68 characters, respectively. Such titles are much easier to understand than the first one with endless sentences (Table 17.3).

Even shorter titles like "Making Common Cause" or "We'll Never Work Alone" reveal context only upon inquiry or during the presentation, thus creating curiosity and interaction.

One of my posters at the time was titled: "Reddening as climatic indicator? Investigations on Quaternary soils and soil sediments of the Balearic Islands" (Fig. 17.1, Wagner et al. 2012). I presented it with great success at two international science conferences, inducing lively discussions with numerous interested colleagues and other experts.

We combine the current topic of climate change with the visual element of soil color. The subtitle more precisely describes what and where we investigated soils. It was by far the most successful poster presentation of my scientific career: there were countless discussions on the topic, all triggered by the short, memorable first sentence. The key factors were the red soil color and the

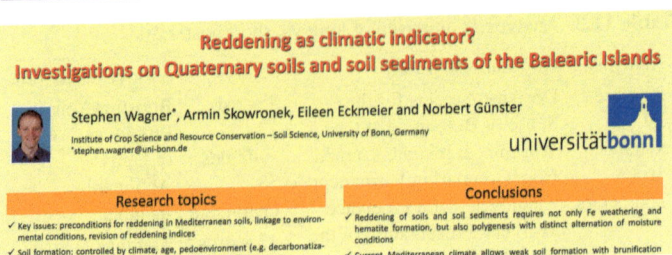

Fig. 17.1 An appealing title of a poster presentation is divided into a provocative question and a thematic subtitle (source: author illustration, poster presented during INQUA Bern 2011, and "Landscapes and Soils Through Time" in Hohenheim 2011, results published in Wagner et al. 2012)

Table 17.4 Titles and subtitles as front covers of journals and magazines

Journal/Magazine	Title: Take Home Message	Subtitle: Key Message
ScienceNews Vol. 205, No. 3 (2024)	Red-Planet Wi-Fi	Connecting on Mars comes with otherworldly challenges
ScienceNews Vol. 204, No. 4 (2023)	The Math of Cake	How to cut a cake fairly is a surprisingly layered problem
ScienceNews Vol. 202, No. 6 (2022)	Unfolding the Secrets of Proteins	Artificial intelligence jump-starts research into how millions of proteins work
Science Vol. 381, Issue 6661 (2023)	Double Trouble	Unraveling Turkey's complex sequence of earthquakes
Nature Vol. 624, Issue 7992 (2023)	One year. Ten stories	10 people who helped shape science in 2023.
Nature Vol. 623, Issue 7988 (2023)	Beat generation	Molecular structure of the myosin filaments that power the hear

visual effect of the poster, coupled with an inviting question. Long after the official end of the session, I was able to engage in lively discussions with other interested parties.

What you can take away from this: use a short title that arouses curiosity. Add a brief subtitle that more precisely leads to your topic. Well-applied examples are available in journals and magazines, such as ScienceNews, Science, and Nature (Table 17.4). All the examples given are illustrated with full-page images.

Chapter 7 introduced Artificial Intelligence (AI) for your Science Pitch. As mentioned, you can use AI tools like ChatGPT as virtual assistants for effective preparation. However, you are responsible for the intellectual work yourself: never copy-paste the wording 1:1, and always check whether the suggested wordings correspond to your style. Your presentation should always remain authentic and personal!

Here, I will share a brief example of how you can develop one sentence pitches using ChatGPT. It is about a research project (GEPRIS 2024) I contributed to after the funding was completed. My last publication on this project was in 2015, so I am simulating the following idea for the scenario in which I could be presenting this project today.

> I wrote this prompt to ChatGPT 4: "Create ten different, attractive, and curiosity-arousing slogans based on the following text from the DFG research project "Paleopedology and Geochronology of the Quaternary on the Balearic Islands". The target group are scientists at a congress. [Here follows the project description stored on the DFG-GEPRIS page (GEPRIS 2024). [translated from German]" ChatGPT then suggests ten ideas for the title of my presentation.

These are alternative presentation titles proposed by Chat-GPT after my prompt:

- Unearth the Secrets of the Balearic Quaternary: A Pedostratigraphic Journey Through Time
- Decoding Climate Mysteries: Dive into the Quaternary Pedostratigraphic Layers
- Reconstructing Western Mediterranean's Climate: A Quaternary Adventure on the Balearics
- Join the Quest: Pioneering Quaternary Climate Research on the Balearic Isles
- From Terra Calcis to Paleosoils: Unraveling the Quaternary Climate Story
- Bridging Epochs: Connecting Central Europe and the Sahara through Quaternary Research
- Discovering the Quaternary Climate Code: A Balearic Expedition Through Time
- Tracing Climate Evolution: Unveiling Quaternary Secrets of the Western Mediterranean
- A Paleoclimatic Puzzle: Assembling the Quaternary History on the Balearic Archipelago
- Pioneering the Past: Crafting a Quaternary Climate Chronicle of the Balearics

I should mention that in earlier prompts within the same chat, I asked ChatGPT to suggest splitting the often-proposed long sentences into a title and a subtitle. ChatGPT seems to have remembered this, as all ten suggestions are divided into two short sentences each. If I like one of the suggestions, I can now adopt it as the title of my next talk. However, I would more likely use parts from one or two suggestions as the basis for a different title, such as "Let's Reconstruct Climate Change in the Western Mediterranean: A Pedostratigraphic Journey Through the Quaternary".

You can do the same for your project by taking the abstract of your publication or conference presentation, the text of your research proposal, or your test speech, transcribing the text, and then asking ChatGPT for suggestions for one sentence pitches.

17.4 A Concise Key Message for Each Part of Your Speech

Enrich your take home message with additional one sentence pitches that you use for each section: the key messages. A concise key message is much better remembered by your audience than a long sentence.

This approach makes sense as soon as you present various facets of your project. Think about individual one sentence pitches for each section. Steve Jobs was a master at this: his key messages have made it into numerous Twitter headlines! In his 2005 Stanford Commencement Address "How to Live Before You Die", the title is the take home message, the one sentence pitch. Jobs articulates clear key messages for the three sections (Fig. 17.2).

They have a sticky effect in a positive sense: they connect the speaker and the audience and motivate them to adopt a new attitude. Jobs repeats some key messages, reinforcing their impact and making them even more memorable. He concludes with a strong final impulse: "Stay hungry. Stay foolish.", repeating this impulse twice more.

Steve Jobs' 2005 Stanford Commencement Address

Take Home Message

How to Live Before You Die: Do What You Love

Part 1	Part 2	Part 3
Connecting the Dots	Love and Loss	Death
Key Messages	**Key Messages**	**Key Messages**
You can only connect the dots looking backward.	The heaviness of being successful vs. the lightness of being a beginner again.	If you live each day as if it was your last, someday you'll most certainly be right.
You have to trust in something.	Have the courage to follow your heart and intuition.	Death is very likely the single best invention of life.
This approach ... has made all the difference.	Keep looking. Don't settle. (2x)	Stay hungry. Stay foolish. (3x)

Fig. 17.2 Division of a speech with a take home message into three sections, each containing key messages. Example: Stanford Commencement Address by Steve Jobs 2005 (source: author illustration)

Jobs underlines the content of his speech with convincing facts and conclusive examples. He frequently adds personal stories and short anecdotes.

Show your passion and ensure that your audience remains attentive and curious. Ideally, they will ask you to tell them more about your project. This will turn the monolog of your Science Pitch into a real dialog.

Answer the following questions:

- What are the key messages behind each section of my speech?
- What will spark my audience's interest in a dialog?

17.5 New Suggestions for Catchy Key Messages

Now you can think about catchy key messages. Two examples from book chapters may further inspire you. They are based on practical, scientifically proven principles.

In the first book, the one sentence pitches of the main chapters (Table 17.5) arouse curiosity. Some of them are concrete statements. Each main chapter is divided into further, very specifically formulated subchapters. I translated the title and each subtitle from German.

The second book has five main parts, each with three to five chapters. These are titled in single words, followed by subtitles with short sentences that reveal more about the content (Table 17.6).

In a nutshell, you should find ways to summarize your content in short single sentences and distinct messages that resonate with your audience (Fig. 17.3).

Table 17.5 One sentence pitches of book chapters in Hüther et al. (2021, pp. 7–8)

Chapter title	Chapter subtitle
Time for Dreams	Starting as an Underdog—Finished as Winner
	A Mad Dream That Came True
	What Team Engagement Has to Do with an Engagement
	Enthusiasm is Fertilizer for the Brain
	Diversity is Better than Simplicity
	The Smaller the Dream, the Quicker It Ends
	Success is a Bittersweet Fruit on the Tree of Achievement

Table 17.6 One sentence pitches of book chapters in Harari (2018, pp. 4–5)

Part	Chapter	Chapter subtitle
Part I. The Technological Challenge	1 Disillusionment	The end of history has been postponed
	2 Work	When you grow up, you might not have a job
	3 Liberty	Big Data is watching you
	4 Equality	Those who own the data own the future
Part II. The Political Challenge	5 Community	Humans have bodies
	6 Civilization	There is just one civilization in the world
	7 Nationalism	Global problems need global answers
	8 Religion	God now serves the nation
	9 Immigration	Some cultures might be better than others
Part III. Despair and Hope	10 Terrorism	Don't panic
	11 War	Never underestimate human stupidity
	12 Humility	You are not the centre of the world
	13 God	Don't take the name of God in vain
	14 Secularism	Acknowledge your shadow

(continued)

Table 17.6 (continued)

Part	Chapter	Chapter subtitle
Part IV. Truth	15 Ignorance	You know less than you think
	16 Justice	Our sense of justice might be out of date
	17 Post-Truth	Some fake news last for ever
	18 Science Fiction	The future is not what you see in the movies
Part V. Resilience	19 Education	Change is the only constant
	20 Meaning	Life is not a story
	21 Meditation	Just observe

Fig. 17.3 Key Takeaway: Your Take Home Message brings essentials to the point (source: author illustration)

References

Asmus, Frank 2021. Impact! Wie Sie sich und andere überzeugen – The Power of Influence. 296 pages, Goldegg.

Asmus, Frank 2024. Der wichtige Unterschied zwischen Kernbotschaft und Schlüsselbotschaften. https://www.linkedin.com/feed/update/urn:li:activity:7163055246863228930?updateEntityUrn=urn%3Ali%3Afs_feedUpdate%3A%28V2%2Curn%3Ali%3Aactivity%3A7163055246863228930%29. Accessed 19 Feb 2024.

Baum, Silke 2020. "Da steht man auf der Straße und bleibt stehen!": eine nutzerinnenorientierte Untersuchung zum Erleben und der Bewältigung von Wohnungsnot bei Frauen. Technische Universität Dresden.

Browning, Jo Filshie 2023. Science Communications matters and how to do it better. TEDx Basel. https://www.youtube.com/watch?v=7Rt8sgt7gNE. Accessed 31 Jan 2024.

GEPRIS, Geförderte Projekte der Deutschen Forschungsgemeinschaft 2024. Paläopedologie und Geochronologie des Quartärs auf den Balearen. DFG-Projektnummer 5174024. https://gepris.dfg.de/gepris/projekt/5174 024?context=projekt&task=showDetail&id=5174024&. Accessed 13 Jan 2024.

Harari, Yuval Noah 2018. 21 Lessons for the 21st Century. 372 pages, Random House.

Hüther, Gerald, Müller, Sven Ole and Bauer, Nicole 2021. Dream-Team. Warum wir nur gemeinsam unser Potential entfalten und unsere Zukunft gestalten können. 288 pages, Goldmann.

Jobs, Steve 2005. How to live before you die. Stanford Commencement Address. https://www.ted.com/talks/steve_jobs_how_to_live_before_ you_die. Accessed 14 Jan 2024.

Pfeiffer, Franziska 2018, 2024. So formulierst du den perfekten Titel deiner Bachelorarbeit. https://www.scribbr.de/aufbau-und-gliederung/titel-bachelorarbeit/. Accessed 19 Feb 2024.

Schlemmer, Kathrin B. 2006. Absolutes und nichtabsolutes Hören. Einflussfaktoren auf das Erinnern von Tonarten. 195 pages, PhD Thesis, Humboldt-Universität zu Berlin. doi: https://doi.org/10.18452/15389. Accessed 19 Feb 2024.

Wagner, Stephen, Skowronek, Armin, Eckmeier, Eileen and Günster, Norbert 2012. Reddening as climatic indicator? Investigations on Quaternary soils and soil sediments of the Balearic Islands. Poster Presentation on INQUA Bern. Quaternary International, 279-280, 524. https://doi.org/10.1016/j. quaint.2012.08.1822. Accessed 14 Jan 2024.

Part III

Science Pitch in Practice

This section provides you with a practical Science Pitch example. Based on this information, you can use the Science Pitch Canvas to summarize your content in an overview. Depending on the purpose and occasion of your Science Pitch, you can use this to prepare your presentation and tailor it to the specific occasion.

Design Your Science Pitch Canvas

Abstract

Use your personal Science Pitch Canvas to prepare your short presentation. The ESPRIT model presented here comprises the content and questions from Chaps. 12 to 17: your expertise and personality, the storyline structure, your performance or appearance, the relevance or significance of your presented topic, your innovation, your USP, and your take home message.

Write down all the answers you consider relevant based on the questions posed in the ESPRIT model. It forms the foundation for the final creation of your Science Pitch presentation!

Now, you have collected all the essential elements for your Science Pitch. Using the resulting canvas, you can now create your individual Science Pitch.

Here is another overview of the main contents of the individual aspects of the ESPRIT (Fig. 18.1).

You can now complete your Science Pitch Canvas independently based on the template of Fig. 18.2. This is your basis for the final creation of your Science Pitch presentation!

© The Author(s), under exclusive license to Springer Fachmedien Wiesbaden GmbH, part of Springer Nature 2024
S. Wagner, *Science Pitch*,
https://doi.org/10.1007/978-3-658-44844-8_18

Science Pitch Canvas: ESPRIT in Your Project

Expertise: Personality

↗ Individual characteristics combined with your professional qualifications.

↗ **Passion before excellence.**

↗ **Your sponsor invests in your personality** first, in the project idea second.

↗ **Solid facts:** Objective attitude, competence, and credibility.

Storyline: Structure

↗ **Vision, Mission, Values:** Personal goal, mission statement, and beliefs.

↗ **Problem and Solution Approach:** Your milestones and distinct findings.

↗ **Storytelling / Hero's Journey** in the project: The story behind your facts and figures in the Narrative Spectrum and Pixar Pitch

↗ **Suspense and Dramaturgy:** Message, audience, facts, and examples

Performance: Presence

↗ **Dramaturgy and Staging:** How do you design your stage performance?

↗ **Authenticity** counts more than perfection

↗ **Performance with passion:** What do you love about your project? What motivates you in particular?

↗ What special personal moments do you share with your audience?

↗ **What do you visualize** on three slides?

Relevance: Significance

↗ **Relevance** of your idea to your field of research and beyond.

↗ **Knowledge transfer:** What is the tangible practical relevance of your project?

↗ **Which industries, companies, and organizations** use this knowledge?

↗ Who reads your publications?

↗ **Feasibility realistic?** Financing, need for time, space, personnel?

Innovation: Your USP

↗ **Pioneering factor:** How innovative, unique, groundbreaking and original is your project?

↗ **Future potential:** Concrete benefit and special added value: For whom?

↗ **Additional revenue** through patent / product?

↗ **Degree of creativity:** Ratio of originality to scientific relevance?

Take Home Message

↗ **One Sentence Pitch:** Your message in the title arouses curiosity

↗ A key message for each speech section

↗ **Sound examples** with convincing facts

↗ **Personal stories and anecdotes**

↗ What do you want your audience to remember most?

Fig. 18.1 Science Pitch Canvas at a glance: all the information that makes up the ESPRIT of your project (source: author illustration)

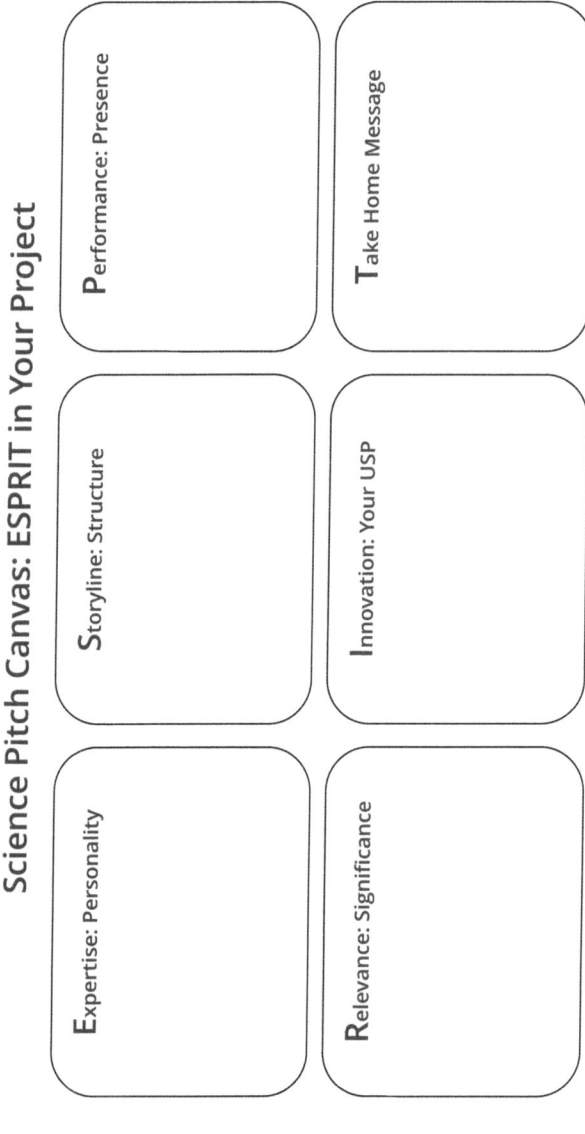

Fig. 18.2 Science Pitch Canvas for completion: prepare your individual Science Pitch (source: author illustration)

Deliver Your Science Pitch

19

Abstract

Once you have finalized your Science Pitch Canvas, you can focus on the setting for your presentation. You put together and prioritize content that fits the occasion and the planned speaking time, tailored to your goal and target audience. Refine your Science Pitch through targeted tests and stay flexible for last-minute changes!

You will get two practical examples of the Science Pitch Canvas. A presentation of the Science Pitch by a PhD student and a brief analysis of this presentation follows this. We also talk about alternatives for a longer talk and a short 1-min talk.

Congratulations! You are now ready to present your Science Pitch on stage!

Here are two examples of how you can create your Science Pitch Canvas. You will get to know two different approaches. You can work with either version or choose to use the bullet point draft (Fig. 19.1) first to prepare the text in the second step. The key is to ensure that these intermediate steps help you to formulate and present your Science Pitch.

In addition to the content outlined in the ESPRIT of Fig. 19.1, you can add further details to your Science Pitch Canvas by answering all the questions in this book, or at least those most relevant to your presentation. If you were presenting your research project at a science conference, your focus would probably be on the methodological part, the most important results, and the conclusions.

Depending on the time available and to suit the occasion and aim of your presentation, you could talk about your passion (in my case, for example, international collaboration) or even add a short anecdote (as in my case, the wine tasting in the research area). On the other hand, the performance part of ESPRIT is mainly limited to background information you consider while preparing and is not part of your wording.

The Science Pitch Canvas in Fig. 19.1 refers to a research project by Wagner (2009) and GEPRIS (2024) on soil formation on marine terraces in Metaponto, Southern Italy. It refers to the fictive preparation of a Science Pitch presentation.

This Science Pitch Canvas (Fig. 19.1) contains details that can be applied to many different occasions, while the final content still needs to be prioritized to design the specific presentation. At a science conference, the focus is usually on the relevance of your project, the innovative part, and your take home message. In contrast, in an interview for a research position, you may prioritize your technical and possibly application-oriented expertise, your work experience, and your values.

Use the ESPRIT model as a flexible basis for future presentations, depending on your purpose, goals, and the allotted speaking time.

The second Science Pitch Canvas (Fig. 19.2) was successfully applied by Christopher Tobe Okolo, Junior Researcher and PhD Student at the Center for Development Research, University of Bonn. His research project is about the "Identification, Distribution and Development of EPN as Biological Control of Insect Pests", with EPN being "entomopathogenic nematodes". In contrast to the first Science Pitch Canvas (Fig. 19.1), Christopher already outlined the potential wording for his Science Pitch delivery (Fig. 19.2). It indicates the speaker preparing for a single event with a particular goal and a specific target audience.

Science Pitch Canvas: Exemplary ESPRIT on a Research Project

Expertise: Personality

- Extensive field and laboratory experience *[list relevant methods]* with research in Australia, Benin, Ecuador, Italy, and Germany; landscape formation through soils and sediments.
- Ambition, perseverance, curiosity *[details and relevant examples, e.g. methodology]*, passion for teamwork and international collaboration.
- Conference presentations in Ecuador, Germany, and New Zealand, 1st publication.

Relevance: Significance

- Spatio-temporal reconstruction of the transgression of coastlines in the Quaternary.
- Basic research in paleopedology to reconstruct landscape history, connection to geomorphology and sedimentology.
- Funding by DFG and a PhD scholarship.
- Publications in Catena, Quaternary International, Revista Mexicana de Ciencias Geológicas. Conference presentations.

Storyline: Structure

- I love the variety of office, field, and laboratory work, thinking, presenting, and publishing.
- Space-time reconstruction by soil chronofunctions. Identify connections between past and current climate changes.
- Soils formed during interglacials AND alluvial sediments accumulated, BUT complex landscape history with increasing soil development, THUS indicating progressive weathering near Metaponto, Southern Italy.

Innovation: Your USP

- Chronological reconstruction of Quaternary morphogenesis of coastal areas in S-Italy.
- Similar soil chronofunctions can be applied to regions with similar parent material and climate.
- Complex interplay of soil-forming processes, paleoclimatic changes, and geomorphological processes prove polygenesis.

Performance: Presence

- Speech at an international science conference with a headset in a lecture hall with 120 seats.
- Business casual clothing. Mental preparation for an enthusiastic performance.
- Highlights: We identified connections between several soil pits. Field work with winetasting (!)
- PowerPoint presentation with 3 slides only, visualizing soil chronosequence, referencing paleopedology projects in the Mediterranean.

Take Home Message

- Paleoclimatic changes and geomorphological dynamics need to be integrated to a greater extent into pedological concepts.
- Our soil chronosequence is the first one to be studied on limestone in the Mediterranean.
- Soil chronofunctions on marine terraces in S-Italy can be applied for pedo-dating.
- Increase in solum thickness and Fe_d/Fe_t ratio are best described by power functions.

Fig. 19.1 Exemplary Science Pitch Canvas for a research project presentation (source: author illustration)

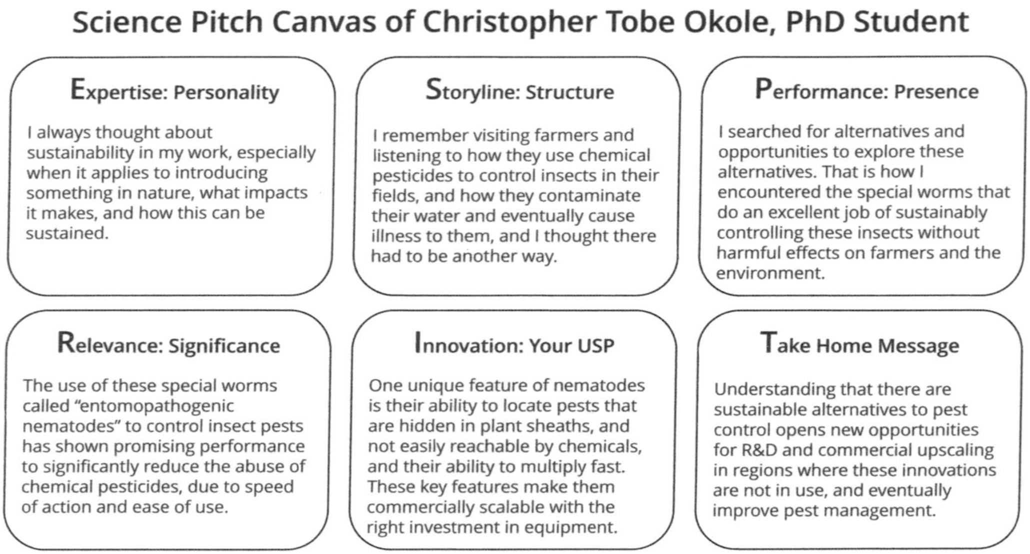

Science Pitch Canvas of Christopher Tobe Okole, PhD Student

Expertise: Personality

I always thought about sustainability in my work, especially when it applies to introducing something in nature, what impacts it makes, and how this can be sustained.

Storyline: Structure

I remember visiting farmers and listening to how they use chemical pesticides to control insects in their fields, and how they contaminate their water and eventually cause illness to them, and I thought there had to be another way.

Performance: Presence

I searched for alternatives and opportunities to explore these alternatives. That is how I encountered the special worms that do an excellent job of sustainably controlling these insects without harmful effects on farmers and the environment.

Relevance: Significance

The use of these special worms called "entomopathogenic nematodes" to control insect pests has shown promising performance to significantly reduce the abuse of chemical pesticides, due to speed of action and ease of use.

Innovation: Your USP

One unique feature of nematodes is their ability to locate pests that are hidden in plant sheaths, and not easily reachable by chemicals, and their ability to multiply fast. These key features make them commercially scalable with the right investment in equipment.

Take Home Message

Understanding that there are sustainable alternatives to pest control opens new opportunities for R&D and commercial upscaling in regions where these innovations are not in use, and eventually improve pest management.

Fig. 19.2 Science Pitch Canvas of Christopher Tobe Okolo for his research project (source: author illustration, with kind permission by Christopher Tobe Okolo)

You now know the essential tools for your Science Pitch. There are just three small steps left to be optimally prepared for your presentation. First, select from your portfolio in the Science Pitch Canvas how you will merge the individual elements into an inspiring orchestra. What ESPRIT do you spread? This is the art of choosing wisely which collected items suit your presentation and inspire your target audience.

Once you have completed your Science Pitch Canvas, prioritize your content. You know the motto "less is more": you do not have to reveal all the information to your audience immediately. Your Science Pitch should arouse curiosity, get people talking about you, and ultimately lead to your project being approved, a job application, or collaboration being accepted. You want to present your expertise with power to the point. Make sure that all six components of the ESPRIT find their place!

As a second step, test your Science Pitch extensively. Record yourself on your smartphone and listen to what you have said. Does everything fit together nicely? You can replace individual sections or rearrange them in a different order. Does the content meet the criteria for the event you are presenting at? Are you making the most of the time available? Now, you can ask friends and colleagues for feedback: where are they particularly attentive and curious? Do they still miss anything? Ultimately, it is your decision what to include in your presentation.

In a third step, record yourself on video: do you radiate enthusiasm? Do you come across as dynamic and engaged? If not, you can simulate your performance further and incorporate any missing aspects. Always remember to be aware of your strengths: self-praise is right!

Now it is about time to deliver your Science Pitch. Here is the 3-min Science Pitch presented by Christopher Tobe Okolo, Junior Researcher and PhD Student at the Center for Development Research, University of Bonn. *[Text slightly edited. I also added my remarks in parentheses.]*

Hello, my name is Christopher Okolo. I am a research scientist in the field of ecology and biological control of pests. I have always have thought about sustainability in my work, especially when it applies to introducing something in nature; what impact does it make, how can this be sustained and what are the aftermath effects? This came into play during my first job as a graduate assistant in a chemical company. I had the opportunity to visit farmers on the field to observe and also get to know how they use chemical pesticides to control pests on crops. *[Expertise; Storyline: Values; Performance: Presence, Passion]* But what really struck me is the use and abuse of these chemicals to an extent that left some residuals in the crops and even killed some non-target organisms. This got me thinking there has to be a way. And honestly, if you do anything, you have to think about the impacts, and alternatives – and yes, ways you can improve on them. *[Storyline: Mission]*

And this, for me, was the beginning of a career searching for alternatives to biological control agents. I found one that caught my attention, which is nematodes: tiny worms that live in the soils, these are called entomopathogenic nematodes. I know it's a mouthful, but I can help you break it down. "Entomo" means insects, "pathogenic" means "causing disease", and "nematodes", these are roundworms. The unique thing about nematodes is that they can reach insects in places where the chemicals cannot reach them, especially in whorls of plant leaves or deep down in the soil. *[Storyline: Mission, Audience Perspective; Performance: Passion]*

They can be multiplied quite easily, and it is possible to mass-produce these nematodes and sell them on a commercial scale. This opens up new investment opportunities in research and development, especially for sending it to areas that are endemically using these chemicals. This, for me, is awesome and an eye-opener for innovative practices, especially in agriculture. *[Relevance; Knowledge Transfer; Innovation: Future Potential]*

I hope you find this quite interesting, and I'm hoping to engage further to discuss this. You should think about sustainability and nematodes. The use of nematodes to control pests is one sustainable pest control measure. *[Take Home Message]*

Thank you!

Christopher delivered his Science Pitch in about 2:40 min, which is close to the optimum of 3 min. The online video highlights a well-prepared performance, passion for the subject as emphasized by a distinct voice, and presence at the moment of the online talk. Even without slides, it is easy to follow through sharing personal insights and translating the critical term "entomopathogenic" into well-understandable parts.

By introducing himself with his expertise, interests, and values, the audience immediately learns about Christopher's background and his close connection to local farmers. The central part of this short talk deals with the mission, taking the audience on a journey and highlighting the innovative part of his research. The relevance of this research project becomes immediately apparent by highlighting the potential of the outcome of this study, mentioning the innovation—or the USP. Christopher also invites further dialog and concludes his Science Pitch with a clear take home message.

Depending on the occasion, he could also deliver his speech as a longer conference presentation or a keynote speech. He would then provide more project details, dive into science specifics, and share personal experiences. In addition, he could visualize the most important and interesting content using a PowerPoint, Keynote, or Prezi presentation.

On the other hand, an even shorter Science Pitch, such as a 1-min talk, would be limited to essentials, consisting of two or three key messages and the final take home message. This shorter version could sound: "Hello, my name is Christopher Okolo. I am a research scientist in the field of ecology and biological control of pests. During my first job as a graduate assistant in a chemical company, I visited farmers observing the use of chemical pesticides to control pests on crops. Since they even killed non-target organisms, I looked for better alternatives and found nematodes. These "entomopathogenic nematodes" are tiny soil worms that reach insects beyond the range of chemicals. It is, therefore, an eye-opener for innovative agriculture practices. This opens up new opportunities for investment in research, development, and sustainable pest control measures."

From my experience, I recommend you prepare at least two or three versions of your presentation. This way, you are well-equipped for last-minute changes. I once prepared a 15-min presentation for an international science congress in Vienna, Austria. As my presentation started late, the moderator did not allow a subsequent discussion, although I had already shortened my presentation by about 3 min. With a clear plan B, I could have prepared my presentation for just 5 min instead of 15. Then, I would most certainly have achieved the main goal of my presentation—a stimulating discussion with an exchange of ideas about my research project.

Always be prepared for last-minute changes!

References

GEPRIS, Geförderte Projekte der Deutschen Forschungsgemeinschaft 2024. Sauer, Daniela. Ableitung von Bodenchronofunktionen auf Meeresterrassen in Süditalien zur Pedo-Datierung. DFG-Projektnummer 41836859. https://gepris.dfg.de/gepris/projekt/41836859. Accessed 14 Jan 2024.

Wagner, Stephen 2009. Soil (Chrono-) Sequences on Marine Terraces. Pedogenesis in two coastal areas of Basilicata and Agrigent, Southern Italy. 327 Seiten, Hohenheimer Bodenkundliche Hefte, Heft 93, PhD Thesis. https://hohpublica.uni-hohenheim.de/bitstreams/7c965f5a-a633-413b-89a7-92170d9d72fa/download. Accessed 16 Nov 2024.

Interview on Science Pitches with Christopher Tobe Okolo (PhD Student)

20

Abstract

According to PhD Student Christopher Tobe Okolo, who delivered a Science Pitch and was interviewed, the ESPRIT model clarifies a story's design and structure. It ensures a focused and understandable presentation of a project's uniqueness and breaks down scientific details into sizable chunks that are understood. The Science Pitch Canvas is a presentation guide covering the uniqueness, personality, research background, and audience benefits. Most times, science ends in a publication. The ESPRIT model challenges us to think more in-depth and highlights potential pitfalls to overcome. The Science Pitch simplifies research into digestible, relatable content beyond complicated research papers. We can spark audience interest with fascinating stories. Passion is crucial and needs to be expressed to and seen by scientific and non-scientific communities. Drama and personality are key elements for a passionate talk.

Scientists seem to talk in a very black box: in numbers that are neither visualized nor translated into stories. Pull out your lab coat and leave the lab, be human, and interact on a very personal level, creating stories out of research. Think about your work as stories and movies. How do you translate and visualize numbers and Excel data sheets into motion pictures?

S. Wagner, *Science Pitch*,
https://doi.org/10.1007/978-3-658-44844-8_20

147

You can express your work by making a movie or playing a Netflix out of your research, making a theater out of your science work.

Think about how you can explain your work in easy-to-understand terms outside science. Create a vision of what you are doing. It spots more interest and creates availability and audience connection.

I interviewed Mr. Christopher Tobe Okolo, Junior Researcher and PhD Student at the Center for Development Research, University of Bonn. Christopher applied the ESPRIT model to develop and deliver his Science Pitch.

Please present your Science Pitch in about 3 min!

Thank you for having me on your interview. Simply put, my interest is to find an alternative to the use of chemicals on the field, instead of using pesticides to control insect pests. From my research, I found out there could be an alternative to the use of biological control agents. And one unique point is called "entomopathogenic nematodes". I know it's a mouthful, but it means nematodes that can control insects that are pathogenic. So, for me, it's an alternative source of controlling insect pests, and of course, makes agricultural products less prone to attack by insects in the field.

[Find the original Science Pitch by Christopher in Chap. 19.]

How would you summarize your topic in one single sentence?

Excessive use of chemicals destroys non-target organisms in the field, but biological control agents can be a source of alternative biological control. Biological control could be an alternative source to control an insect.

If there was only one aspect of your research, what should your audience remember after your Science Pitch? How do you want to be remembered?

Well, just know that not everything you find under the soil can be harmful to your plants. There are beautiful creatures under the soil that are beneficial to the plant, and this is what nematodes, the entomopathogenic pathogenic nematodes can do.

What aspect of the ESPRIT model in the Science Pitch is most valuable to you? What surprised you?

There is a lot of research done to identify new samples from different regions. For me, the unique one was to find a sample they have never been discovered in Nigeria before and which wasn't really in the mainstream. But most importantly it performed well, and I don't want to go into the scientific details of this, but for me, the key selling point is that each region holds a unique mystery and a unique solution to inherent problem-solving.

What I loved about the ESPRIT model was how it gave clarity to designing my story and to the structure of what I had to say. I could easily have said this in a very disorganized way, but the ESPRIT model helped me to focus on each of the components in a well-tailored manner. And that way the audience can follow the story and understand the uniqueness of my project.

Is there any additional element or idea you would add to the ESPRIT model?

The ESPRIT model, I think, is unique, and I don't know how you came about this, but there was lots of thinking into this. This, for me, is perfect. Maybe one thing I could add is where you talked about the performance and personal story. I think it also adds to the uniqueness of the story as well. I think there might be an overlap, but on the whole, the ESPRIT model is unique. If this is your model, I think you can also patent it.

How did your Science Pitch help you to work on your presentation?

It is said that scientists are not good communicators because there are so many details about what we do in a scientific enclave. But speaking to the public who have no idea of all the details becomes a challenge. What the ESPRIT model helped me to do is to break down what I have done into very sizable chunks that can be understandable. Following this model makes it quite easy to relate with other people.

The Science Pitch Canvas provides us with a good platform to itemize in very clear detail what is the uniqueness of each component, how the personality comes up, what are the back ends of the research, and who can benefit from it. These are very clear points that might be lost if I were to give the presentation without any guide, and it might be so disorganized. This is one uniqueness of the ESPRIT model: giving clarity to each point and helping me to understand which other details I need to emphasize, to give that clarity to the audience.

What are the challenges you had to deal with to get to the point successfully?

It takes a lot of thinking because there are some points that I had to think about more in-depth and how to fit into the ESPRIT model. I give you an example: thinking about the uniqueness, and how to build it into another project. Most times, science ends in the publication, and then you go on to do another experiment.

But thinking about who will benefit from it, how do I build this into a different project, and how do I expand the reach? That is one component that not all science projects think about. With the ESPRIT model, looking at that component gives me the challenge to think more in-depth. For me, it was highlighting some of the pitfalls that I have to overcome while designing my Science Pitch.

Do you see a difference between science publications and presentations? What should scientists learn to present better, compared to publications?

A while ago, I had a presentation at the "Global Forum for Food and Agriculture", where I presented at the Science Slam, which for me was a new stage to present my work to a non-scientific audience. It was a lot of work for those 10 min. The research paper is already very complicated, and not everybody can read it and understand it. But breaking it down to a very sizable, understandable, and relatable form is something that we can do more often so that we break the walls of science or take science out of the box, which is one of your key mantras.

I love how some unique scientists have done that, like Neil deGrasse Tyson, who I consider a good scientific communicator. I think more people need to do that more often to get more people into their research but in a more relatable way.

At which events and networks do you pitch outside the Science Slam? What is your experience with science presentations and with science pitching?

Normally in the scientific world, we present at scientific conferences. And occasions rarely occur where we have to present outside the science in a community. The Science Slam was unique, and I would say this is one big event where I have made a presentation that does not consist of the science community or most of the science community.

One other platform is presenting at clubs like Toastmasters, which is not a scientific community but provides a platform to relate with other people and share my ideas on my research work with a non-scientific community. I haven't explored wide opportunities, but these two platforms, the Science Slam and then presenting to other Toastmasters, are the two non-scientific communities where I've presented my work. I wish I could do more of that. If you have an idea where I can also share my presentation on my results, I would love to. For me, that is more of networking and getting people the insight of what I'm doing.

Who is your major target group presenting your science project? What are your goals while delivering your talk and project to your target group? What questions are you asked there, depending on the target groups?

The primary beneficiaries of my work are the farmers in the field, but most importantly, the people who will educate and enlighten the farmers, and then the higher echelon, the people who will make these available.

There are three layers: the primary end users, the connectors to other people who are going to educate the farmers, and then the people at the very top who would make this available, those who would sponsor it. When you look at these three layers, you realize that they have different objectives.

My main goal is to fit in the objective of these three different layers. For the guys at the ministry, at the government level who will finance it, it's to understand that there are alternatives to whatever they have on their hands, and they need to pay key attention to this new aspect. And for the guys who will enlighten the farmers, it's a source of excitement that they don't have to deal with one single component. For the farmers, it's a relief instead of using chemicals that are harmful to the environment and human beings, they can also learn something new.

For me, achieving these three main goals is quite task-intense, but it also fuels the passion to continue doing what I'm doing.

What else inspired your audience or your target group?

I think this is the point where I can also share why I got into this research in the first place, which I shared in the Science Pitch. I started out working in a chemical company. I had first-hand interaction with farmers using these insecticides, and I had first-hand experience with how the abuse of these chemicals was impacting the people and how it was destroying soil life and the wider environmental factors.

On a large scale, it was impacting negatively on climate change because these chemicals degrade into gases that accumulate greenhouse gases which deplete the ozone layer. This is more like the wider picture. Bringing all this together, I thought: "No, there has to be an alternative!" That was when I came across the use of alternative means to control insects. Fortunately, I got sponsorship to pursue this at the Master's and PhD levels. Thinking about building this into a project, into a business, and reaching out to a wider community on the relevance and the beauty is that it has been done in other societies. My case is to extend it to other regions that have not experienced it and have no idea what this is all about. For me, this is one key goal.

What new opportunities open up once you present yourself and your project? What has changed for you since then?

There are lots of interests that build up when I present this or when I speak to people. The number one thing that comes up is: "How do we get them? Your story is so fascinating. This seems like a life changer. I don't like chemicals: where do we get them?" This kind of feedback gives me joy, it gives me fulfillment that what I'm doing has an impact.

But I have also chosen another challenge: how do I get this out? Not just the information, but the products. It's one thing to have information to tell people there is an alternative. It's another thing to say, "Okay, this is it." That is the stage where I'm heading to. Building a product, establishing it so that people can use it. I get lots of feedback and stories where people tell me, "Oh, plants were attacked by insects, and I tried everything, but the chemicals are failing. Where can we get your products?"

For now, there is no scaling product. That is the next point that I'm getting to. For me, it's progress. And the feedback inspires the passion to follow this up to the next level.

What similarities do you see between Science Pitches and presentations in general? What distinguishes the Science Pitch from your perspective?

When I am passionate about what I am doing, it shows—irrespective of the platform, be it on a Science Slam, presenting to a non-science community, or being at a scientific conference: how I express my passion for it speaks out loud. For me, that is the similarity. The key differences are the ways I present it, the words that I use, the description that I use, and the technicalities of my presentation.

But what I believe is that the passion has to be there. It has to be seen in my face, in the way I demonstrate it. As you can see right now, we can see I'm so passionate about what I do. This has to be shown clearly, irrespective of the community, be it a non-scientific or a scientific community, dramatizing and showing the personality of the passion that I have for what I'm doing. These are the key elements that stand out, irrespective of the platform.

In your experience, what are scientists' strengths when presenting themselves and their projects? What could be improved in science communication and presentation?

I will start with the weaknesses because it's quite evident. After all, scientists seem to talk in a very black box, again referencing your slogan. We tend to speak just in numbers without trying to add a picture or trying to visualize those numbers and make meaning out of them. This is also more evident when you listen to scientists who have made a good impact but have lost or have no idea how to immerse themselves into demonstrating this in a very visual manner.

So, for me, the key weakness is leaving the lab, being human, interacting with people on a very personal level, and creating stories out of what they do. One key strength that I can identify in a scientist is that he/she can overcome these weaknesses, can build stories out of what they say because there are so many beautiful and exciting moments, especially in the laboratories, on the field, or in other scientific settings, but translating this into stories becomes difficult.

Some scientists have been able to acknowledge this and try to build stories. The strength is to build on those weaknesses and overcome them, transforming this into a very, very exciting activity. And always look forward to sharing the stories, leaving the enclaves of those very tiny black boxes, and pulling out the lab coats to become a human, and relate to other people on a very personal level.

What advice can you share with other scientists and experts when they present their project?

Think about your work as stories. Create movies out of your work. Whatever you do in your scientific laboratories, or on the field, or whatever, think about it in terms of stories; think about how these numbers, your data, how the things on the Excel sheet can be visualized: how these can be translated into motion pictures. Create a movie and play a Netflix in your head while doing your laboratory work. And think about you sitting down in the theater and watching a movie about what you are doing. I think that way, it will improve how scientists express their work.

But more importantly, it will increase their passion; it will help them relate more than what they do like, for instance, when I'm in the laboratory with nematodes, I'm working there, and I look at nematodes, and I sing for them, and I tell stories with the nematode. I like to relate as I was talking to humans. For me, it's a way to build a passion, but also dramatize what I'm doing. It makes it easier for me to relate with a non-scientist and also to spot more interest when people get into science. This is one thing I will give to other scientists: make a movie out of your science work, play a Netflix out of your science work. Go to the theater and make a theater out of your science work.

Is there anything else you want to share with your audience that we have not yet covered? What is your take home message?

I think we covered most of it. I would just reiterate some of the key points that I mentioned, and that is trying to express your personality when you are presenting your work and seeking opportunities to present outside of the scientific world. This has helped me to immerse myself in science, but also while doing the science, think about what I'm doing.

How does the non-scientist benefit? How can I translate what I'm doing to someone else outside my field, outside the field of science? How do I tell my story to someone who is studying philosophy, someone who is studying the arts? How do I create that vision of what I'm doing? I think this way it kind of spurs more interest in what we do in the lab, but also creates more availability for ourselves, to express ourselves outside the laboratories.

Thank you very much.

Thank you very much for having me. I appreciate it.

Outro: Bring Your Expertise with Power to the Point

21

Abstract

Now you have all the information you need for presenting your research and getting to the point with your Science Pitch. With growing experience, your presentations will become more fluent and confident. The points discussed in this book are your guide, a framework for your best possible preparation. Be sure to consider the criteria for the event you are presenting at. Above all, have fun with what you are presenting. With expertise and enthusiasm about your project, you will find a close connection with your audience!

Imagine raising your voice for 3 min to talk about your research project. The audience includes professors, scientists, and students, as well as a committee of research sponsors and donors, potential project partners, and company representatives, altogether with interested members of the public.

You will talk about your vision, how your project will raise the standards of knowledge in your field, and why you and your team are the right fit due to experience and expertise. With a specific practical example, you translate your project's technical content, facts, and milestones into a captivating narrative. This captures the attention of your audience and sparks their interest. You are truly immersed and deliver a passionate talk.

© The Author(s), under exclusive license to Springer Fachmedien Wiesbaden GmbH, part of Springer Nature 2024
S. Wagner, *Science Pitch*,
https://doi.org/10.1007/978-3-658-44844-8_21

You look ahead to the future, clarifying the innovation and uniqueness of your project. In your final statement, you succinctly summarize the entire project once again. Then you receive your well-deserved applause for a successful Science Pitch.

Now you have all the information you need for presenting your research and getting to the point with your Science Pitch. With growing experience, your presentations will become more fluent and confident. The points discussed in this book are your guide, a framework for your best possible preparation. Be sure to consider the criteria for the event you are presenting at. Above all, have fun with what you are presenting. With expertise and enthusiasm about your project, you will find a close connection with your audience!

Imagine raising your voice for 3 min to talk about your research project. The audience includes professors, scientists, and students, as well as a committee of research sponsors and donors, potential project partners, and company representatives, altogether with interested members of the public.

You will talk about your vision, how your project will raise the standards of knowledge in your field, and why you and your team are the right fit due to experience and expertise. With a specific practical example, you translate your project's technical content, facts, and milestones into a captivating narrative. This captures the attention of your audience and sparks their interest. You are truly immersed and deliver a passionate talk.

You look ahead to the future, clarifying the innovation and uniqueness of your project. In your final statement, you succinctly summarize the entire project once again. Then, you receive your well-deserved applause for a successful Science Pitch.

I wish you the best possible success whenever you present your Science Pitch to an audience!

If you are interested in further information about the Science Pitch or want to connect, you will find it all under the two links below:

https://www.DrStephenWagner.com/SciencePitch/
https://www.linkedin.com/in/DrStephenWagner/

Dr. Stephen Wagner